Practical HTML and CSS

Elevate your internet presence by creating modern and high-performance websites for the web

Brett Jephson

Lewis Coulson

Ana Carolina Silveira

Practical HTML and CSS

Group Product Manager: Kaustubh Manglurkar
Publishing Product Manager: Urvi Sambhav Shah
Book Project Manager: Srinidhi Ram
Senior Content Development Editor: Feza Shaikh
Technical Editor: Simran Ali
Copy Editor: Safis Editing
Indexer: Rekha Nair
Production Designer: Aparna Bhagat
Marketing Coordinators: Nivedita Pandey

First published: November 2019

Second edition: November 2024

Production reference: 1260924

Published by Packt Publishing Ltd.
Grosvenor House
11 St Paul's Square
Birmingham
B3 1RB, UK.

ISBN 978-1-83508-091-7

www.packtpub.com

To my daughters, Adeline and Viola, for keeping me busy, having too much fun, and making it hard to find time to write anything.

– Brett Jephson

To Rebecca, Louis, Dylan, and Mae, for all of your love and support.

– Lewis Coulson

To my sister, Angelica, who bravely raised me surrounded by books.

– Ana Carolina Silveira

Contributors

About the authors

Brett Jephson is a product engineer specializing in frontend engineering and design systems. He has worked in media, games, and app companies for many years, putting all the latest and greatest web technologies to use in many interesting ways while championing accessibility and good UI.

I would like to thank Kate for supporting me and doing more than her share of looking after the kids (Adeline and Viola) while I worked on this book.

Lewis Coulson is a senior frontend developer with over 10 years of commercial experience. He has worked at a range of organizations ranging from creative agencies to government departments.

I would like to thank my family who have supported me whilst writing this book: Rebecca, Louis, Dylan, and Mae.

Ana Carolina Silveira is a seasoned frontend web developer, known for her expertise in CSS and her deep understanding of user experience and behavior. As an open source enthusiast, she actively participates in community projects, sharing her knowledge and passion for web development with others.

First and foremost, I would like to thank God for granting me the knowledge needed for this project; SDG. I am also deeply grateful to my husband, Lucas, for his support and encouragement throughout the process of writing this book. A heartfelt thanks as well to my family, friends.

About the reviewer

Akhilesh Tiwari is a software engineer with over 20 years of experience. He obtained his MTech from BITS Pilani, India. Akhilesh has developed software solutions for the world's leading organizations, such as Merck, Novartis, BNY Mellon, Fujitsu, Cognizant, and Persistent Systems to name a few. Akhilesh now resides in New Jersey, USA, where he enjoys a fulfilling family life with his wonderful wife and two adorable children.

First of all, I would like to thank the Supreme Personality of Godhead, Krishna, for everything. I would like to thank my loving parents for their unconditional love and support. I would not have been able to do this book review without the support of my loving wife, Sanchita, and my two young ones, Osh and Radha. Last but not least, I would like to thank Packt for this opportunity and their trust.

Table of Contents

Part 1: Introducing HTML and CSS

1

Introduction to HTML and CSS   3

2

Structure and Layout   43

3

Text and Typography Styling   83

Part 2: Understanding Website Fundamentals

4

Creating and Styling Forms   127

5

Adding Animation to Web Pages   171

6

Themes, Color, and Polishing Techniques 201

Part 3: Building for All

7

Using CSS and HTML to Boost Performance 229

8

Responsive Web Design and Media Queries 249

9

Ensuring Accessibility in HTML and CSS 283

Part 4: Advanced Concepts

10

SEO Essentials for Web Developers 319

11

Preprocessors and Tooling for Efficient Development   357

12

Strategies for Maintaining CSS Code   393

13

The Future of HTML and CSS – Advancements and Trends   425

Preface

With knowledge of HTML and CSS, you can build visually appealing, interactive websites without relying on pre-packaged website-building tools and their limitations. *Practical HTML and CSS* takes you on a journey, showing you how to create beautiful websites using your own content while understanding how they work and how to manage them in the long term.

The book begins by introducing HTML5 and CSS3, guiding you through the website development process with easy-to-follow steps. You'll explore how browsers render websites, progressing to adding rich multimedia elements such as video and audio to create a more cinematic experience. You'll also use JavaScript to add interactivity to your site, build intuitive forms to capture user data, and add slick transitions with animations. You'll delve into mobile-first development using responsive design and media queries, ensuring your sites perform seamlessly across all devices.

Throughout, you'll engage in hands-on projects and learn how to craft beautiful, responsive applications. Each chapter builds on the last, expanding your knowledge and equipping you with the skills needed to craft stunning, professional websites. Along the way, you'll also learn how to optimize your sites for performance and SEO, preparing you to tackle the most in-demand technologies in the industry.

Whether you're a beginner or looking to refine your expertise, this guide serves as a comprehensive roadmap for your journey and a trusted companion throughout your career. As you progress, you'll find yourself returning to its pages for advice, inspiration, and best practices. With *Practical HTML and CSS*, you're not just learning a set of tools—you're building a solid foundation that will support your growth as a web developer for years to come.

Who this book is for

The two technologies covered in this book, HTML and CSS, are available to anyone with a web browser and a text editor. This means this book is aimed at anyone who wants to put together a web page and is looking for some practical exercises to help them learn. Here's who will benefit most from this guide:

- **Aspiring web developers**: If you're just starting your journey in web development, this book provides a solid foundation in HTML and CSS. You'll gain a comprehensive understanding of these essential technologies, allowing you to build your first websites with confidence.

- **Junior developers**: For those with some experience in web development, this guide offers deeper insights into best practices and advanced techniques. It will help you enhance your skill set, improve your portfolio, and advance in your career.

- **UI/UX designers**: Understanding the structure and styling of web pages is crucial for designers. This book will enable you to better collaborate with developers by understanding how your designs are implemented using HTML and CSS.

- **Freelancers and entrepreneurs**: If you're building or managing your own website, this guide will equip you with the knowledge to create and maintain your web presence without relying on external developers.

The book is accessible to someone learning web development for the first time but it delves into a lot of subject areas and goes beyond the fundamentals of what you can do with HTML and CSS. We hope, therefore, that there is something useful for anyone in their formative years as a web developer.

What this book covers

Chapter 1, Introduction to HTML and CSS, provides a grounding in the fundamentals of web development. Through practical exercises, the chapter introduces you to the syntax of both HTML and CSS, looks at the relationship between HTML and CSS, and explains the roles of these technologies in rendering a web page in the browser.

Chapter 2, Structure and Layout, introduces you to the structural elements in HTML, including header, footer, and section tags. You will also learn the three main CSS layout techniques: float, flex, and grid.

Chapter 3, Text and Typography Styling, introduces you to text-based elements, such as paragraphs, headings, and lists. We will go over how to style text-based elements in web pages and let your creativity run wild.

Chapter 4, Creating and Styling Forms, introduces you to the creation of web forms, starting with the key HTML elements used in forms and then learning how to style them with CSS. Along the way, you will also learn how to style your forms with validation styling.

Chapter 5, Adding Animation to Web Pages, explores how to effectively incorporate animations into web applications using CSS. You'll be introduced to the versatile `transform` property, the intricacies of z-index, the smooth transitions enabled by CSS, and the powerful `@keyframes` rule for creating complex animations. By the end of this chapter, you'll have created a portfolio-worthy project that demonstrates how these techniques can be applied in various real-world scenarios.

Chapter 6, Themes, Color, and Polishing Techniques, dives into managing styles based on user preferences using CSS themes and other refining techniques. You'll gain an understanding of color theory with HSL, and how to implement light and dark themes, utilize CSS filters, and apply effective typography. We'll also explore how to combine these approaches to enhance and personalize the user experience, ensuring your web designs are both visually appealing and user-centric.

Chapter 7, Using CSS and HTML to Boost Performance, provides an understanding of how we can measure web performance using Core Web Vitals. Through practical exercises, it explains some simple techniques using HTML and CSS that we can use to improve the performance of a web page.

Chapter 8, Responsive Web Design and Media Queries, teaches you how to implement responsive applications using breakpoints and media queries. We'll delve into the "mobile first" approach, ensuring that your applications are optimized for mobile devices without sacrificing performance or design quality. You'll also learn about viewports, responsive values, and the incredibly useful Flexbox layout, equipping you to create adaptive, high-quality web applications.

Chapter 9, Ensuring Accessibility in HTML and CSS, explains some of the challenges that web pages can present to users with disabilities and provides simple techniques using HTML and CSS that we can employ to improve the experience for those users while also improving the experience for all users.

Chapter 10, SEO Essentials for Web Developers, provides an in-depth guide to understanding SEO and how to measure its performance. You'll learn about Core Web Vitals, strategies to improve SEO metrics, and how to position your application for high visibility in search engines. This chapter focuses on mastering the tools and techniques to outsmart search algorithms, giving your web projects a competitive edge.

Chapter 11, Preprocessors and Tooling for Efficient Development, takes a step further into performance and scalability. You'll learn how to architect and build scalable, maintainable applications using powerful build tools such as Gulp and Webpack to automate development processes. We'll also explore preprocessors such as Less and Sass, which enhance the capabilities of CSS, making your development process more efficient and your code more manageable.

Chapter 12, Strategies for Maintaining CSS Code, focuses on CSS performance and maintainability, introducing methodologies such as **BEM** (which stands for **Block, Element, Modifier**) and tools such as Tailwind CSS. These approaches help you maintain clean, organized, and efficient CSS code, ensuring your projects remain scalable and easy to manage as they grow.

Chapter 13, The Future of HTML and CSS – Advancements and Trends, provides a look at the latest advancements and emerging trends in HTML and CSS. This chapter serves as a guide to the future of web development, offering insights into the technologies and features that will shape tomorrow's web. As the final chapter of this book, this chapter opens the door to the cutting-edge resources and innovations you can start exploring today to stay ahead in the rapidly evolving world of web development.

To get the most out of this book

Software/hardware covered in the book	Operating system requirements
HTML5	Any OS with a web browser
CSS	Any OS with a web browser
Chrome Dev Tools	Any OS with Chrome web browser

If you are using the digital version of this book, we advise you to type the code yourself or access the code from the book's GitHub repository (a link is available in the next section). Doing so will help you avoid any potential errors related to the copying and pasting of code.

Download the example code files

You can download the example code files for this book from GitHub at `https://github.com/PacktPublishing/Practical-HTML-and-CSS-Second-Edition`. If there's an update to the code, it will be updated in the GitHub repository.

We also have other code bundles from our rich catalog of books and videos available at `https://github.com/PacktPublishing/`. Check them out!

Conventions used

There are a number of text conventions used throughout this book.

Code in text: Indicates code words in text, database table names, folder names, filenames, file extensions, pathnames, dummy URLs, user input, and Twitter handles. Here is an example: "To do this, create a folder named `styles` inside the project's folder."

A block of code is set as follows:

```css
/* Layout styles */
.container {
  max-width: 1200px;
  margin: 0 auto;
  padding: 0 20px;
}

.grid {
  display: flex;
  flex-wrap: wrap;
}
```

When we wish to draw your attention to a particular part of a code block, the relevant lines or items are set in bold:

```html
<html lang="en">
<head>
  <meta charset="UTF-8">
  <meta
    name="viewport"
    content="width=device-width, initial-scale=1.0"
  >
  <link rel="stylesheet" href="styles/base.css">
  <link rel="stylesheet" href="styles/layout.css">
  <link rel="stylesheet" href="styles/module.css">
  <link rel="stylesheet" href="styles/state.css">
  <link rel="stylesheet" href="styles/theme.css">
</head>
```

Bold: Indicates a new term, an important word, or words that you see onscreen. For instance, words in menus or dialog boxes appear in bold. Here is an example: "As this is an example taken from the Packt website, you will notice that it contains items such as the company logo, search bar, and the **Sign In** button."

> **Tips or important notes**
> Appear like this.

Get in touch

Feedback from our readers is always welcome.

General feedback: If you have questions about any aspect of this book, email us at customercare@packtpub.com and mention the book title in the subject of your message.

Errata: Although we have taken every care to ensure the accuracy of our content, mistakes do happen. If you have found a mistake in this book, we would be grateful if you would report this to us. Please visit www.packtpub.com/support/errata and fill in the form.

Piracy: If you come across any illegal copies of our works in any form on the internet, we would be grateful if you would provide us with the location address or website name. Please contact us at copyright@packt.com with a link to the material.

If you are interested in becoming an author: If there is a topic that you have expertise in and you are interested in either writing or contributing to a book, please visit authors.packtpub.com.

Share Your Thoughts

Once you've read *Practical HTML and CSS*, we'd love to hear your thoughts! Scan the QR code below to go straight to the Amazon review page for this book and share your feedback.

https://packt.link/r/183508091X

Your review is important to us and the tech community and will help us make sure we're delivering excellent quality content.

Download a free PDF copy of this book

Thanks for purchasing this book!

Do you like to read on the go but are unable to carry your print books everywhere?

Is your eBook purchase not compatible with the device of your choice?

Don't worry, now with every Packt book you get a DRM-free PDF version of that book at no cost.

Read anywhere, any place, on any device. Search, copy, and paste code from your favorite technical books directly into your application.

The perks don't stop there, you can get exclusive access to discounts, newsletters, and great free content in your inbox daily

Follow these simple steps to get the benefits:

1. Scan the QR code or visit the link below

https://packt.link/free-ebook/978-1-83508-091-7

2. Submit your proof of purchase
3. That's it! We'll send your free PDF and other benefits to your email directly

Part 1:
Introducing HTML and CSS

In this part, we'll explore the fundamental principles of creating web pages. We'll start by learning about the fundamental concepts behind how web pages work. Next, we'll explore techniques for structuring our web pages and styling them with custom layouts. Finally, we'll delve into text and typography, paying close attention to styling concerns.

This section contains the following chapters:

- *Chapter 1, Introduction to HTML and CSS*
- *Chapter 2, Structure and Layout*
- *Chapter 3, Text and Typography Styling*

1
Introduction to HTML and CSS

Whether you want to build a web page to advertise your business, blog about a hobby, or maintain an online community, **HyperText Markup Language** (HTML) and **Cascading Style Sheets** (CSS) are the foundations upon which you will build.

HTML and CSS work together but each has a different role in making a web page. A **web page** is made up of lots of different types of content (text, images, links, video, and audio). HTML structures that content and **CSS** styles it. Together, they tell the browser how and what to render.

These two technologies are simple to get started with and provide enough power and expressivity that they let you get your ideas out there to the vast audience on the web.

Navigate to a website and what you see is the rendered output of content marked up with HTML and styled with CSS. As a browser user, you have access to the source code of a web page. In Chrome, for example, you can view a page's source code by pressing the keys *Ctrl + U* on a PC or *command + option + U* on a Mac. Alternatively, you can right-click with a mouse and choose **View Page Source**. Try it yourself. As an example, the following two figures show what the Packt website's **Web Development** portal looks like when rendered in the browser and as source code respectively.

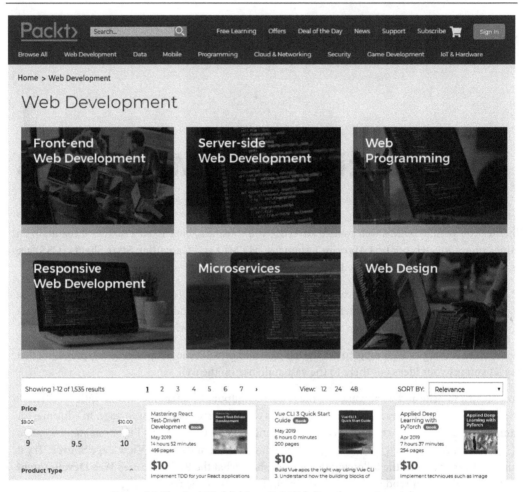

Figure 1.1: The Packt Publishing site's Web Development portal

aHR0cHM6Ly93d3cucGFja3RwdWIuY29tL2dldC1yZXdhcmRzL2xvZ2luL300

Figure 1.2: The HTML source code of the Packt site

By learning to write the HTML and CSS that make up the source code, we can create our own modern website.

In this chapter, we will develop an understanding of how a web page renders by following the process from initial request to completed composition. We will learn how to build a simple web page using HTML and then we will learn how to style that page with CSS.

The following topics will be covered in the chapter:

- How a web page renders
- Understanding HTML
- Understanding CSS

By the end of this chapter, you will have been introduced to two of the core technologies of the web – HTML and CSS – and you will understand their roles in creating websites. You will have created a page from scratch and you will have used selectors to target parts of that web page for styling.

Technical requirements

The code files for this chapter can be found at `https://packt.link/808iP`.

How a web page renders

When we navigate to a web page in our favorite browser, what happens? How do HTML and CSS translate to the page we see in front of us? In other words, how does a web page render?

The following figure shows a flowchart of the process, which is then further explained:

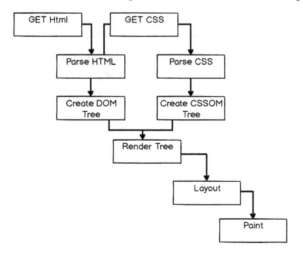

Figure 1.3: Flow chart of the web page render process

To summarize the process:

1. The user starts by navigating to a URL, possibly via a hyperlink or by typing the URL into the address bar of their browser.

2. The browser will make a GET request to the disk or a network. It will read the raw bytes from that location and convert them to characters (based on character encoding, such as UTF-8).

3. The browser then parses these characters according to the HTML standard to find the tokens that are familiar as HTML elements, such as <html> and <body>.

4. Another **parse** is then made to take these tokens and construct objects with their properties based on the rules appropriate to that token. At this point, the objects are defined.

5. Finally, the browser can define the relationships between these objects and construct the HTML DOM (Document Object Model) for the web page.

6. At this point, we have a DOM but not a rendered web page. The next task is to construct the **CSSOM** (CSS Object Model). Again, the browser will load any style sheet resources it needs to, which were found while parsing the document. It will then construct the styles associated with nodes in the tree structure, which gives us the CSSOM.

7. With the information gathered in the DOM and the CSSOM, the browser can create a render tree. The render tree is constructed by combining information from the CSSOM and the HTML DOM. Nodes in the HTML DOM that will not be rendered (for instance, those with the display: none; style) are excluded from the render tree. Those that are rendered are enriched with their computed style rules.

8. Now the browser has all the information it needs, it can begin to calculate the positions of elements in the rendered viewport. This is called the **layout stage**. The browser lays elements out based on their size and position within the browser viewport. This stage is often also called **reflow**. It means the browser must recalculate the positions of elements in the viewport when elements are added to or removed from the page or when the viewport size is changed.

9. Finally, the browser will rasterize or paint each element on the page, depending on their styles, shadows, and filters to render the page the user will see.

In this section, we have given a brief and simplified summary of the rendering of a web page. Think about how many resources might be loaded on a relatively complicated website and with JavaScript running events and we can see that much of this process happens frequently and not in such a linear manner. We can start to see the complexities of what a browser is doing when it renders your web page.

In the next section, we will start to look at how we create a web page by learning about the syntax of HTML, which helps us structure and contextualize our content. We will learn about the syntax of HTML and the elements we can use, and we will apply this knowledge to create a simple web page.

Understanding HTML

HTML is a markup language used to describe the structure of a web page.

Consider a snippet of text with no markup:

```
HTML HyperText Markup Language (HTML) is a markup language used to
describe the structure of a web page. We can use it to differentiate
such content as headings lists links images Want to https://www.
packtpub.com/web-development Learn more about web development.
```

The preceding snippet of text makes some sense. It may also raise some questions. Why does the snippet begin with the word HTML? Why is there a URL in the middle of a sentence? Is this one paragraph?

Using HTML, we can differentiate several bits of content to give them greater meaning. We could mark the word HTML as a heading, `<h1>HTML</h1>`, or we could mark a link to another web page using `<a href="https://www.packtpub.com/web-development">Learn more about web development</a>`.

There have been several versions of HTML since its first release in 1993 at the beginning of the web. Throughout the rest of this chapter, and indeed the rest of this book, we will be looking at and working with the current version of the HTML language, HTML5, which is the 5th major version of HTML. When we use the term HTML, we will refer specifically to HTML5 and if we need to talk about a different version we will do so explicitly (e.g., HTML 4.01).

In the next section, we will look at the syntax of HTML in more detail.

Syntax

The syntax of HTML is made up of **tags** (with angle brackets, `<>`) and attributes. HTML provides a set of tags that can be used to mark the beginning and end of a bit of content. The opening tag, closing tag, and all content within those bounds represent an HTML element. The following figures show the HTML element representation without and with tag attributes respectively:

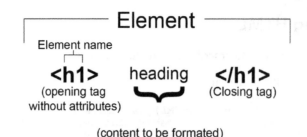

Figure 1.4: HTML element representation without tag attributes

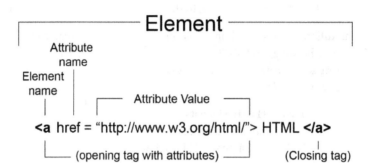

Figure 1.5: HTML element representation with tag attributes

A tag has a name (for instance, p, img, or h1), and that name combined with attributes will describe how the browser should handle the content. Many tags have a start and end tag with some content in between, but some tags don't expect any content, and these can be self-closing.

An opening tag can have any number of **attributes** associated with it. These are modifiers of the element. An attribute is a name-value pair. For example, href="https://www.packtpub.com/web-development" is an attribute with the name of href and the value of https://www.packtpub.com/web-development. An href attribute represents a hypertext reference or a URL, and when this attribute is added to an anchor element, <a>, it creates a hyperlink that the user can click in the browser to navigate to that URL.

To provide information within an HTML document to be ignored by the parser and not shown to the end user, you can add **comments**. These are useful for notes and documentation to aid anyone who might read or amend the source of the HTML document. A comment begins with <!-- and ends with -->. Comments, in HTML, can be single or multiline. The following are some examples:

```
<!-- Comment on a single line -->

<!--
   This comment is over multiple lines.
```

```
  Comments can be used to inform and for detailed
  documentation.
-->
```

You can use comments to provide helpful hints to other developers working on the web page but they will be ignored by the browser when parsing the page.

Let's see what the previous snippet of text content looks like when it is given some meaning with HTML:

```
<h1>HTML</h1>
<p>
  HyperText Markup Language (HTML) is a markup language
  used to describe the structure of a web page.
</p>
<p>
  We can use it to differentiate such content as:
</p>
<ul>
  <li>headings</li>
  <li>lists</li>
  <li>links</li>
  <li>images</li>
</ul>
<p>
  Want to <a href="https://www.packtpub.com/web-
  development">learn more about web development?</a>
</p>
```

If we were to look at this HTML code rendered in a browser, it would look like the following figure:

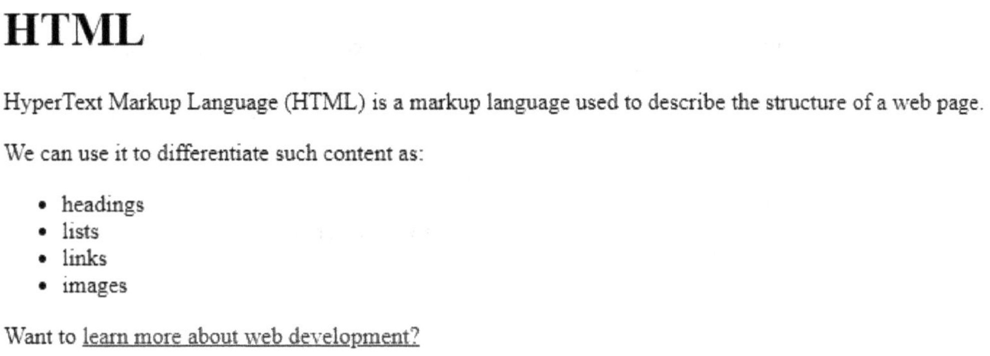

HTML

HyperText Markup Language (HTML) is a markup language used to describe the structure of a web page.

We can use it to differentiate such content as:

- headings
- lists
- links
- images

Want to learn more about web development?

Figure 1.6: HTML rendered in the Google Chrome web browser

The first line shows the HTML text content with a start tag, <h1>, and an end tag, </h1>. This tells the browser to treat the text content as an h1 heading element.

The next line of our code snippet has a `<p>` start tag, which means the content until the corresponding end tag, `</p>` (on the last line), will be treated as a **paragraph** element. We then have another paragraph and then an unordered list element that starts with the `<ul>` start tag and ends with the `</ul>` end tag. The unordered list has four child elements, which are all list item elements (from the `<li>` start tag to the `</li>` end tag).

The last element in the example is another paragraph element, which combines text content and an anchor element. The anchor element, starting from the `<a>` start tag and ending at the `</a>` end tag, has the `learn more about web development?` text content and an `href` attribute. The `href` attribute turns the anchor element into a hyperlink, which a user can click to navigate to the URL given as the value of the `href` attribute.

As with our example, the contents of a paragraph element might be text but can also be other HTML elements, such as an anchor tag, `<a>`. The relationship between the anchor and paragraph elements is a parent-child relationship.

HTML elements

HTML5 defines more than a hundred tags that we can use to mark up parts of an HTML document. These include the following:

- **The document root element**: `<html>`
- **Metadata elements**: `<base>`, `<head>`, `<link>`, `<meta>`, `<style>`, and `<title>`
- **Content sectioning elements**: `<address>`, `<article>`, `<aside>`, `<body>`, `<footer>`, `<header>`, `<h1>`, `<h2>`, `<h3>`, `<h4>`, `<h5>`, `<h6>`, `<main>`, `<nav>`, and `<section>`
- **Block text elements**: `<blockquote>`, `<dd>`, `<details>`, `<dialog>`, `<div>`, `<dl>`, `<dt>`, `<figcaption>`, `<figure>`, `<hr>`, `<li>`, `<menu>`, `<ol>`, `<p>`, `<pre>`, `<summary>`, and `<ul>`
- **Inline text elements**: `<a>`, `<abbr>`, `<b>`, `<bdi>`, `<bdo>`, `<br>`, `<cite>` , `<code>`, `<data>`, `<dfn>`, `<em>`, `<i>`, `<kbd>`, `<mark>`, `<q>`, `<rp>`, `<rt>`, `<ruby>`, `<s>`, `<samp>`, `<small>`, `<span>`, `<strong>`, `<sub>`, `<sup>`, `<del>`, `<ins>`, `<time>`, `<u>`, `<var>`, and `<wbr>`
- **Media elements**: `<area>`, `<audio>`, `<img>`, `<canvas>`, `<map>`, `<track>`, `<video>`, `<embed>`, `<iframe>`, `<object>`, `<picture>`, `<portal>`, `<source>`, `<svg>`, and `<math>`
- **Scripting elements**: `<noscript>` and `<script>`
- **Table elements**: `<caption>`, `<col>`, `<colgroup>`, `<table>`, `<tbody>`, `<td>`, `<tfoot>`, `<th>`, `<thead>`, and `<tr>`

- **Form elements**: `<button>`, `<datalist>`, `<fieldset>`, `<form>`, `<input>`, `<label>`, `<legend>`, `<meter>`, `<optgroup>`, `<option>`, `<output>`, `<progress>`, `<select>`, and `<textarea>`

- **Web component elements**: `<slot>` and `<template>`

We don't have to know all of these tags to use HTML well; some fulfill more common use cases than others. Each has a distinct purpose and provides a different semantic meaning and throughout this book, we will go into some detail about how to use these elements.

Content types

When starting with HTML, it can be easy to find the number and variety of elements overwhelming. It may be helpful to think about HTML in terms of content types.

The following table has a description and example of the different content types that can describe an element:

Type	Description	Example
Metadata	Content hosted in the head of an HTML document. Doesn't appear in the web page directly but is used to describe a web page and its relationship to other external resources.	`<meta name="viewport" content="width=device-width,initial-scale=1.0">`

Flow	Text and all elements that can appear as content in the body of an HTML document.	`<body>`
		
		`<h1>Heading</h1>`
		
		`<p>Some content...</p>`
		
		`</body>`

Sectioning	Used to structure the content of a web page and to help with layout. Elements in this category are described in Chapter 2, Structure and Layout.	`<aside></aside>`
		
		`<article class="blog-post">`
		
		`  <section></section>`
		
		`</article>`

Phrasing	Elements such as those used for marking up content within a paragraph element. Chapter 3, Text and Typography, will be largely concerned with this content type.	`<p><em>Emphasized text</em> and some normal text.</p>`

Type	Description	Example

Heading	Elements used to define the headings of a section of an HTML document. The h1-6 elements represent headings with h1 having the highest ranking.	`<h1>Main Heading</h1>`
		
		`<h2>Subheading</h2>`

Embedded	Embedded content includes media, such as video, audio, and images.	`<img src="media/kitten.png" alt="A cute kitten">`
Interactive	Elements that a user can interact with, which include media elements with controls, form inputs, buttons, and links.	`<input type="password" name="password" required>`

Table 1.1: Different content types

Let's run through an example of how an element can fit into these category types using the `<img>` element.

If we want to embed an image in our web page, the simplest way is to use the img element. If we want to create an img element, an example of the code looks like this: `<img src="media/kitten.png" alt="A cute kitten">`.

We set the src attribute on the img element to an image URL; this is the source of the image that will be embedded in the web page.

Unless your image has no value other than as a decoration, it is a very good idea to include an alt attribute. The alt attribute provides an alternative description of the image as text, which can then be used by screen readers if an image does not load, or in a non-graphical browser.

> **Note**
>
> A **screen reader** is a software application that allows people who are visually impaired or blind to access and interact with a computer. The screen reader allows a user to navigate a web page with a keyboard and will output the content as speech. We will look further at accessibility in *Chapter 9*.

An img element is a form of **embedded** content because it embeds an image in an HTML document. It can appear in the body of an HTML document as the child element of the body element, so it would be categorized as **flow** content.

An image can be included as content in a paragraph, so it is a type of **phrasing** content. For example, we could have inline images appear in the flow of a paragraph:

```
<p>
   Kittens are everywhere on the internet. The best thing
   about kittens is that they are cute. Look here's a kitten
   now:
   <img src="media/kitten.jpg" alt="A cute kitten">.
   See, cute isn't it?
</p>
```

This code would render the following figure, with the image embedded in the paragraph and the rest of the text flowing around it:

Kittens are everywhere on the internet. The best thing about kittens is that they are

cute. Look here's a kitten now: . See, cute isn't it?

Figure 1.7: Image with text flowing around it

In certain circumstances, an `img` element is a type of **interactive** content. For this to be the case, the image must have a `usemap` attribute. The `usemap` attribute allows you to specify an image map, which defines areas of an image that are treated as hyperlinks. This makes the image interactive.

An `img` element does not act as metadata and it does not provide a sectioning structure to an HTML document. Nor is it a heading.

Elements can appear in more than one category and there is some overlap between the relationships of the categories. Some of these elements are very common and are used often, but some of these elements have very specific purposes and you may never come across a use case for them.

The content types can be useful for understanding how elements work together and which elements are valid in where. For further reference, we can see where each available element is categorized in the W3C's documentation on HTML5: `https://html.spec.whatwg.org/multipage/dom.html#kinds-of-content`.

The HTML document

A web page is made up of an HTML document. The document represents a hierarchical tree structure similar to a family tree. Starting from a **root** element, the relationship between an element and its contents can be seen as that of a parent element and a child element. An element that is at the same level of the hierarchy as another element is a sibling to that element. We can describe elements within a branch of the tree as **ancestors** and **descendants**.

This structure can be represented as a tree diagram to get a better idea of the relationship between elements.

Take, for example, this simple HTML document:

```
<html>
  <head>
    <title>HTML Document structure</title>
  </head>
  <body>
    <div>
      <h1>Heading</h1>
      <p>First paragraph of text.</p>
      <p>Second paragraph of text.</p>
    </div>
  </body>
</html>
```

Here, the root is an html element. It has two children: a head element (containing a title) and a body element containing some more content. It can be represented as a tree diagram as follows:

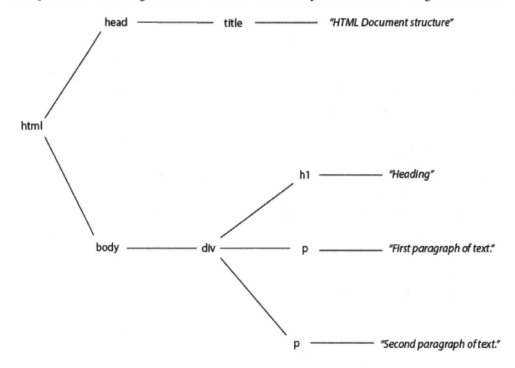

Figure 1.8: A representation of the HTML document as a tree diagram

In the browser, this code would render the following web page:

Figure 1.9: HTML rendered in the Google Chrome web browser

The `<html>` element is the parent of both the `<head>` and `<body>`, which (as children of the same parent) are siblings. `<body>` has one child, a `<div>` tag, and that has three children: an `<h1>` element and two `<p>` elements. The `<h1>` element is a descendant of `<body>` but not of `<head>`.

Understanding this structure will become more important when we look at CSS selectors and how we target parts of the HTML document later in this chapter.

Structuring an HTML document

An HTML5 document normally starts with a doctype declaration and has a root `html` element with two children – the `head` element and the `body` element.

The doctype declaration tells the browser it is dealing with an HTML5 document. The doctype is `<!DOCTYPE html>` and appears as the first line of the document. It is recommended to always add a doctype to make sure your HTML document renders as expected.

> **Note**
>
> The doctype declaration is not case sensitive, so variations such as `<!doctype html>` and `<!DOCTYPE HTML>` are equally valid.

One of the nice things about HTML5 is that it simplifies doctype declaration. Before HTML5, there were two commonly used variations of web markup – HTML4 and XHTML1 – and they both had strict, transitional, and frameset versions of their doctype declarations. For example, the HTML 4 strict declaration looked like this: `<!DOCTYPE HTML PUBLIC "-//W3C//DTD HTML 4.01// EN" "http://www.w3.org/TR/html4/strict.dtd">`.

After the doctype, we have the `html` element, which is the root of the HTML document.

It is strongly recommended that you add a `lang` attribute to your `html` element to allow the browser, screen readers, and other technologies, such as translation tools, to better understand the text content of your web page.

The two children of the `html` element are as follows:

- The `head` element, which includes the `title` and metadata providing information about assets to load and how web crawlers and search engines should handle the page.

- The `body` element, which mostly represents the content rendered for a human browser user to consume. This includes articles, images, and navigation.

  In code, the structure we have described looks like this:

  ```
  <!doctype html>
  <html lang="en">
    <head><title>Page Title</title></head>
    <body></body>
  </html>
  ```

This code would result in a blank web page with no content or metadata.

Metadata

The `head` is home to most machine-read information in an HTML document. The browser, screen readers, and web crawlers can get a lot of information from metadata and handle the web page differently depending on that information.

The following elements are considered metadata content:

- `base`: This lets you set a base URL

- `link`: This determines the relationship between a page and a resource (such as an external style sheet)

- `meta`: This a catch-all for metadata

- `title`: This is the name of your web page as it appears in the browser tab and search results and is announced by screen readers

- The `meta` element can represent many different types of metadata, including some used by social networks to represent a web page.

Some common usages include the following:

- Setting character encoding for a page – `<meta charset="utf-8">`
- Setting the viewport for a browser on a mobile device – `<meta name="viewport" content="width=device-width, initial-scale=1">`

These elements give web developers ways to tell a browser how to handle the HTML document and how it relates to its environment. We can describe our web page for other interested parties (such as search engines and web crawlers) using metadata.

Our first web page

In our first example, we will create a very simple web page. This will help us to understand the structure of an HTML document and where we put different types of content.

Exercise 1.01 – creating a web page

In this exercise, we will create our first web page. This will be the minimal foundation upon which future chapters can build.

> **Note**
> The complete code for this exercise can be found at `https://packt.link/SduQx`.

The steps are as follows:

1. To start, we want to create a new folder, `chapter_1`, and then open that folder in Visual Studio Code (**File | Open Folder...**).
2. Next, we will create a new plain text file and save it as `index.html`.
3. In `index.html`, we start by adding the doctype declaration for HTML5:

   ```
   <!DOCTYPE html>
   ```

4. Next, we add an HTML tag (the root element of the HTML document):

   ```
   <html lang="en">
   </html>
   ```

5. In between the opening and closing tags of the html element, we add a head tag. This is where we can put metadata content. For now, the head tag will contain a title:

```
<head>
    <title>HTML and CSS</title>
</head>
```

6. Below the head tag and above the closing html tag, we can then add a body tag. This is where we will put the majority of our content. For now, we will render a heading and a paragraph:

```
<body>
    <h1>HTML and CSS</h1>
    <p>
        How to create a modern, responsive website
        with HTML and CSS
    </p>
</body>
```

The result of this exercise should look like the following figure when opened in a browser:

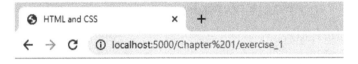

HTML and CSS

How to create a modern, responsive website with HTML and CSS

Figure 1.10: The web page as displayed in the Chrome web browser

Activity 1.01 – video store page template

We've been tasked with creating a website for an online on-demand film store called *Films On Demand*. We don't have designs yet but want to set up a web page boilerplate that we can use for all the pages on the site.

We will use comments as placeholders to know what needs to change for each page that is built on top of the boilerplate template. For visible content in the body element, we will use *lorem ipsum* to get an idea of how content will flow.

The steps are as follows:

1. Create a file named `template.html`.

2. We want the page to be a valid HTML5 document. So, we will need to add:

    *   The correct doctype definition.

    *   **Elements to structure the document**: The `html` element, the `head` element, and the `body` element.

    *   A `title` element that combines the *Films on Demand* brand with some specifics about the current page.

    *   **Metadata to describe the site**: We'll set this to `Buy films from our great selection. Watch movies on demand.`

    *   Metadata for the page character set and a `viewport` tag to help make the site render better on mobile browsers.

3. We want to add placeholders for a heading (an `h1` element) for the page, which we will populate with *lorem ipsum*, and a paragraph for the content flow, which we will also populate with the following *lorem ipsum* text:

    `"Lorem ipsum dolor sit amet, consectetur adipiscing elit. Nullam quis scelerisque mauris. Curabitur aliquam ligula in erat placerat finibus. Mauris leo neque, malesuada et augue at, consectetur rhoncus libero. Suspendisse vitae dictum dolor. Vestibulum hendrerit iaculis ipsum, ac ornare ligula. Vestibulum efficitur mattis urna vitae ultrices. Nunc condimentum blandit tellus ut mattis. Morbi eget gravida leo. Mauris ornare lorem a mattis ultricies. Nullam convallis tincidunt nunc, eget rhoncus nulla tincidunt sed. Nulla consequat tellus lectus, in porta nulla facilisis eu. Donec bibendum nisi felis, sit amet cursus nisl suscipit ut. Pellentesque bibendum id libero at cursus. Donec ac viverra tellus. Proin sed dolor quis justo convallis auctor sit amet nec orci. Orci varius natoque penatibus et magnis dis parturient montes, nascetur ridiculus mus."`

> **Note**
> The solution to this activity can be found at `https://packt.link/WbEPx`

In this section, we've looked at HTML, the markup language that structures and gives context to the content of a web page. We have looked at the syntax of HTML, created our first web page, and learned about the structure of an HTML document. When we've looked at our web page in a browser, it has been rendered with the default styling provided by the browser. In the next section, we will look at how we can customize the styling of our web page using **CSS**. We will learn how to add styles, how to specify what parts of a page they apply to, and some of the properties we can style.

Understanding CSS

CSS is a style sheet language used to describe the presentation of a web page.

The language is designed to separate concerns. It allows the design, layout, and presentation of a web page to be defined separately from content semantics and structure. This separation keeps source code readable and lets a designer update styles separately from a developer who might create the page structure or a web editor who is changing content on a page.

A set of CSS rules in a style sheet determines how an HTML document is displayed to the user. It can determine whether elements in the document are rendered, how they are laid out on the web page, and their aesthetic appearance.

In the next section, we will look at the syntax of CSS.

Syntax

A **CSS declaration** is made of two parts: a property and a value. The property is the name for some aspect of style you want to change; the value is what you want to set it to.

Here is an example of a CSS declaration:

```
color: red;
```

The property is `color` and the value is `red`. In CSS, `color` is the property name for the foreground `color` value of an element. That essentially means the color of the text and any text decoration (such as underline or strikethrough). It also sets a `currentcolor` value.

For this declaration to have any effect on an HTML document, it must be applied to one or more elements in the document. We do this with a selector. For example, you can select all the <p> elements in a web page with the p selector. So, if you wanted to make the color of all text in all paragraph elements red, you would use the following CSS ruleset:

```
p {
  color: red;
}
```

The result of this CSS ruleset applied to an HTML document can be seen in the following figure:

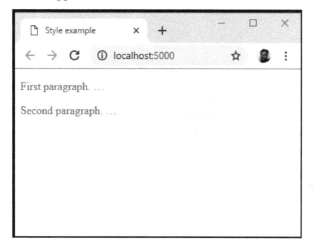

Figure 1.11: Result of a CSS rule applied to <p> elements in HTML

The curly braces represent a declaration block and that means more than one CSS declaration can be added to this block. If you wanted to make the text in all paragraph elements red, bold, and underlined, you could do that with the following ruleset:

```
p {
    color: red;
    font-weight: bold;
    text-decoration: underline;
}
```

The result of this CSS ruleset applied to an HTML document can be seen in the following figure:

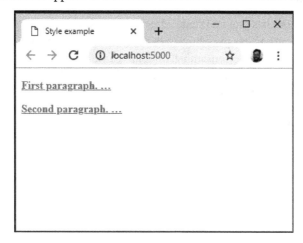

Figure 1.12: Several CSS declarations applied to <p> elements in HTML

Multiple selectors can share a CSS ruleset. We can target these with a comma-separated list. For example, to apply the color red to p elements, h1 elements, and h2 elements, we could use the following ruleset:

```
p, h1, h2 {
   color: red;
}
```

Multiple CSS rulesets form a style sheet. The order of these CSS rules in a style sheet is very important as this is partly how the cascade or specificity of a rule is determined. A more specific rule will be ranked higher than a less specific rule and a higher-ranked rule will be the style shown to the end user. We will look at cascade and specificity later in this chapter:

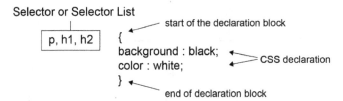

Figure 1.13: A CSS ruleset explained

Adding styles to a web page

There are several ways we can add CSS to a web page:

- Via an element's style attribute (inline styles)
- With a style element in the head or body of the HTML document
- By linking an external style sheet to the HTML document with a link element
- Each of these methods has pros and cons. Inline styles apply only to that element and have high specificity but we can't access pseudo-classes and pseudo-elements this way. They can make maintenance and updates time-consuming. Both the link and style elements provide greater separation of concerns, keeping the CSS separated from the HTML, which can be beneficial for organization and maintaining a clean code base.

We will try out each of these methods in the following exercises.

Exercise 1.02 – adding styles

In this exercise, we will add styles to a web page using the link element, the style element, and the style attribute.

> **Note**
>
> The complete code for this exercise can be found at `https://packt.link/yspKi`.

Here are the steps:

1. Let's start with a simple web page:

```html
<!DOCTYPE html>
<html lang="en">
    <head>
        <meta charset="utf-8">
        <title>Adding styles</title>
    </head>
    <body>
        <h1>Adding styles</h1>
        <p>First paragraph</p>
        <p>Second paragraph</p>
    </body>
</html>
```

Before adding any styles, the web page will look like this:

Figure 1.14: The unstyled web page

2. We'll make the text more readable by creating an external CSS file that we can link to our web page. To do that we save a file named `styles.css` in the `chapter_1` folder.

3. Add the following CSS to `styles.css`:

```css
body {
    font-family: Arial, Helvetica, sans-serif;
    font-size: 18px;
    padding: 0;
    margin: 1rem;
}

h1 {
    margin: 0;
    margin-bottom: 1rem;
}

p {
    margin: 0;
    margin-bottom: .5rem;
}
```

4. Next, we need to link the file to the web page. We do this with a `link` element added to the tag `<head>` of the web page beneath the tag `<title>`:

```html
<link href="styles.css" rel="stylesheet">
```

5. Using a `style` element, added just before the end tag of `body`, we can set a different color for all paragraph elements:

```html
<style>
p {
    color: red;
}
</style>
```

The result will look similar to the following figure, with all paragraphs colored red:

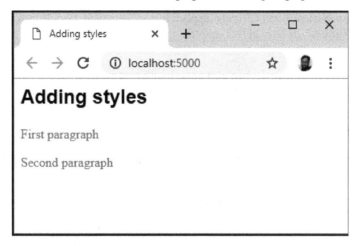

Figure 1.15: The web page with styles applied

6. Finally, we will give the first paragraph a different style using an inline style attribute, setting the color to blue and adding a line-through text decoration as follows:

```
<p style="color: blue; text-decoration: line-through">
    First paragraph
</p>
```

The result will be like the one shown in the following figure:

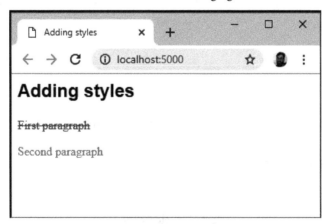

Figure 1.16: The web page with inline style applied

> **Note**
> Inline styles have precedence over CSS rules applied in a style element. We will look at specificity and the rules of the cascade later in this chapter.

We've looked at how we can add styles to a web page. In the next section, we will look at the different CSS selectors we can use to apply our styles to parts of the HTML document.

CSS selectors

To target elements in the HTML document with CSS, we use **selectors**. There are a lot of options available to help you select a wide range of elements or very specific elements in certain states.

Selectors are a powerful tool and we will look at them in some detail as the different options available can help with both web page performance and making your CSS more maintainable.

For example, you can use a selector to target the first letter of a heading, like you might expect to see in a medieval book:

```css
h1::first-letter {
    font-size: 5rem;
}
```

Or, you could use a selector to invert the colors of every odd paragraph in an article:

```css
p {
    color: white;
    background-color: black;
}
p:nth-of-type(odd) {
    color: black;
    background-color: white;
}
```

We will explore a variety of the options available to us when creating selectors.

Element, ID, and class

Three commonly used selectors are as follows:

- **Element type**: For example, to select all p elements in an HTML document, we use the p selector in a CSS ruleset. Other examples are h1, ul, and div.

- **A class attribute**: The `class` selector starts with a dot. For example, given the `<h1 class="heading">Heading</h1>` HTML snippet, you could target that element with the `.heading` selector. Other examples are `.post` and `.sub-heading`.

- **An ID attribute**: The id selector starts with a hash symbol. For example, given the `<div id="login"> <!-- login content --> </div>` HTML snippet, you could target this element with the `#login` selector. Other examples include `#page-footer` and `#site-logo`.

The universal selector (*)

To select all elements throughout an HTML document, you can use the universal selector, which is the asterisk symbol (`*`). Here is an example snippet of CSS that is often added to web pages; a value is set on the `html` element and then inherited by all descendant elements:

```
html {
    box-sizing: border-box;
}
*, *:before, *:after {
    box-sizing: inherit;
}
```

Using the `inherit` keyword and the universal selector, we can pass a value on to all the descendants of the `html` element. This snippet will universally apply the border-box model to all elements and their pseudo-elements (that's the reason for `:before` and `:after`). You'll learn more about the box model and layout in the next chapter.

Attribute selectors

Attribute selectors let you select elements based on the presence of an attribute or based on the value of an attribute. The syntax is square brackets, `[]`, with the suitable attribute inside. There are several variations that you can use to make matches:

- `[attribute]` will select all elements with an attribute present; for example, `[href]` will select all elements with an `href` attribute.

- `[attribute=value]` will select all elements with an attribute with an exact value; for example, `[lang="en"]` will select all elements with a `lang` attribute set to en.

- `[attribute^=value]` will select all elements with an attribute with a value that begins with the matching value; for example, `[href^="https://"]` will select all elements with an `href` attribute beginning with `https://`, which links to a secure URL.

- `[attribute$=value]` will select elements with an attribute with a value that ends with the matching value; for example, `[href$=".com"]` will select all elements with an `href` attribute that ends with `.com`.

- `[attribute*=value]` will select elements with an attribute with a value that has a match somewhere in the string; for example, `[href*="co.uk"]` will select all elements with an `href` attribute matching `.co.uk`. `http://www.example.co.uk?test=true` would be a match, as would `https://www.example.co.uk`.

Pseudo-classes

To select an element when it is in a particular state, we have several pseudo-classes defined. The syntax of a pseudo-class is a colon, `:`, followed by a keyword.

There are a great number of pseudo-classes, but most developers' first experience of them is when styling links. A link has several states associated with it:

- When an anchor element has an `href` attribute, it will have the `:link` pseudo-class applied to it

- When a user hovers over the link, the `:hover` pseudo-class is applied to it

- When the link has been visited, it has the `:visited` pseudo-class applied to it

- When the link is being clicked, it has the `:active` pseudo-class applied to it

Here is an example of applying styling to the various pseudo-class states of an anchor element:

```
a:link, a:visited {
  color: deepskyblue;
  text-decoration: none;
}

a:hover, a:active {
  color: hotpink;
  text-decoration: dashed underline;
}
```

In the following figure, we can see the first link with the `:link` or `:visited` styles applied and the second link with the `:hover` or `:active` styles applied:

Figure 1.17: Link with and without the hover state

The cascade can cause some issues with styling links. The order in which you specify your CSS rules for each state of the link is important. If, for example, we applied the a:hover rule before the a:link rule in the previous example, we would not see the hover effect. A mnemonic exists for remembering the order: *love-hate*. The *l* is for :link, the *v* is for :visited, the *h* is for :hover, and the *a* is for :active.

Some other useful pseudo-classes for selecting elements in a particular interactive state include :checked, :disabled, and :focus.

Several pseudo-classes help us select a pattern of children nested under an element. These include :first-child, :last-child, :nth-child, :nth-last-child, :first-of-type, :last-of-type, :nth-of-type, and :nth-last-of-type.

For example, we can use :nth-child with an unordered list to give a different style to list items based on their position in the list:

```
<style>
  ul {
    font-family: Arial, Helvetica, sans-serif;
    margin: 0;
    padding: 0;
  }

  li {
    display: block;
    padding: 16px;
  }
  li:nth-child(3n-1) {
    background: skyblue;
    color: white;
    font-weight: bold;
```

```
    }

  li:nth-child(3n) {
    background: deepskyblue;
    color: white;
    font-weight: bolder;
  }
</style>

<!-- unordered list in HTML document -->
<ul>
  <li>Item 1</li>
  <li>Item 2</li>
  <li>Item 3</li>
  <li>Item 4</li>
  <li>Item 5</li>
  <li>Item 6</li>
  <li>Item 7</li>
</ul>
```

The following figure shows the result. The :nth-child pseudo-class gives you a lot of flexibility because you can use keywords such as odd and even or functional notation such as 3n - 1:

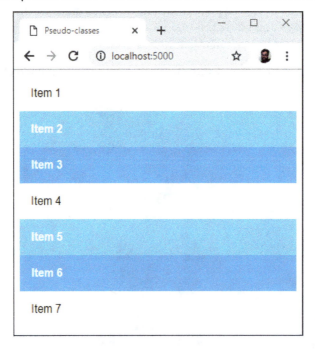

Figure 1.18: Using the :nth-child pseudo-class

Pseudo-elements

Pseudo-element selectors are preceded by two colons (::) and they are used to select part of an element. The available pseudo-elements include ::after, ::before, ::first-letter, ::first-line, ::selection, and ::backdrop.

These pseudo-elements give us a handle we can use to add stylistic elements without adding to the HTML document. This can be a good thing if the pseudo-element has no semantic value and is purely presentational, but it should be used with care.

Combining selectors

What makes CSS selectors particularly powerful is that we can combine them in several ways to refine our selections. For example, we can select a subset of li elements in an HTML document that also has a .primary class selector with li.primary.

We also have several options, sometimes called **combinators**, for making selections based on the relationships of elements:

- To select all the li elements that are descendants of an ul element, we could use ul li.

- To select all the li elements that are direct children of an ul element with the primary class, we might use ul.primary > li. This would select only the direct children of ul.primary and not any li elements that are nested.

- To select a li element that is the next sibling of li elements with the selected class, we could use li.selected + li.

- To select all of the li elements that are the next siblings of li elements with the selected class, we could use li.selected ~ li.

The following figure shows the difference between using li.selected + li and li.selected ~ li. In other words, the difference between the following two CSS declarations is applied to a list where the fourth list item has a .selected class applied to it:

```
li.selected + li {
  background: deepskyblue;
  color: white;
  font-weight: bolder;
}
li.selected ~ li {
  background: deepskyblue;
  color: white;
```

```
        font-weight: bolder;
    }
```

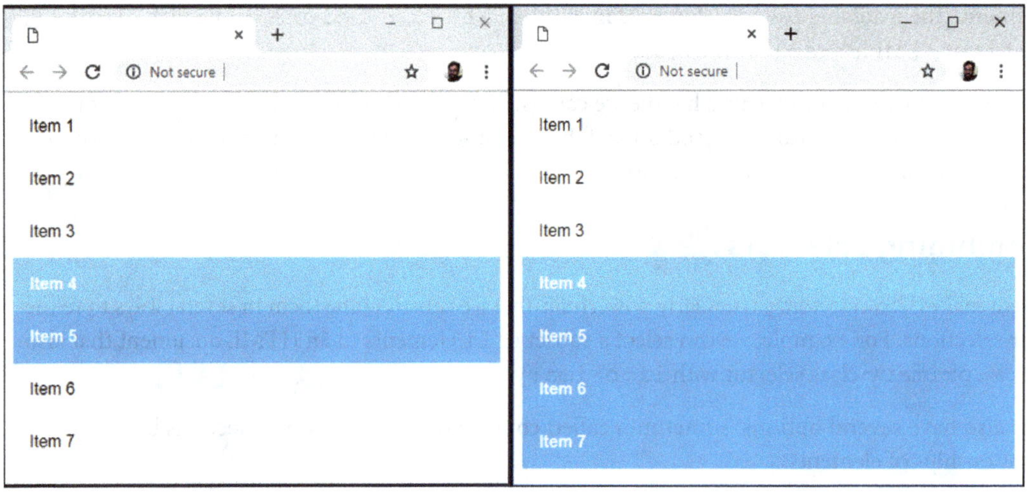

Figure 1.19: Selecting the next adjacent sibling compared to selecting all of the next siblings

Let's try out some of the selectors we've learned about in an exercise.

Exercise 1.03 – selecting elements

In this exercise, we will differentiate list items by styling the odd items. We will use a class selector to style a selected item and a next-siblings combinator to style the elements after the selected item.

> **Note**
> The complete code for this exercise can be found at https://packt.link/KaThH.

The steps are as follows:

1. We will start with a simple web page with a ul list element and nine list items:

    ```
    <!DOCTYPE html>
    <html lang="en">
        <head>
            <meta charset="utf-8">
            <title>Selectors</title>
        </head>
    ```

```
    <body>
        <ul>
            <li>Item 1</li>
            <li>Item 2</li>
            <li>Item 3</li>
            <li>Item 4</li>
            <li>Item 5</li>
            <li>Item 6</li>
            <li>Item 7</li>
            <li>Item 8</li>
            <li>Item 9</li>
        </ul>
    </body>
</html>
```

2. So that we can style a selected item differently, we will add a `selected` class to the fifth list item:

    ```
    <li class="selected">Item 5</li>
    ```

3. Next, we will add a `style` element to the `head` element with the following CSS:

    ```
    <head>
        <meta charset="utf-8">
        <title>Selectors</title>
        <style>
            ul {
                font-family: Arial,
                    Helvetica, sans-serif;
                margin: 0;
                padding: 0;
            }

            li {
                display: block;
                padding: 16px;
            }
        </style>
    </head>
    ```

This will remove some of the default styling of the unordered list in the browser. It will remove margins and padding on the list and set the font style to `Arial` (with `Helvetica` and `sans-serif` as a fallback).

4. Next, we will style the odd list items with the :nth-child pseudo-class. We can use the odd keyword for this. With this style, any odd list item will have a blue background and white text:

```
li:nth-child(odd) {
    background-color: deepskyblue;
    color: white;
    font-weight: bold;
}
```

This gives us the stripy effect that we can see in the following figure:

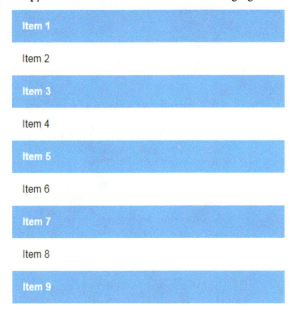

Figure 1.20: Stripy list using :nth-child(odd)

5. We can style the selected class selector:

```
li.selected {
    background-color: hotpink;
}
```

This overrides the striped effect for those items with the `selected` class selector, as seen in the following figure:

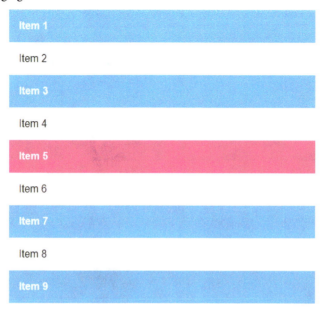

Figure 1.21: Stripy list with a selected item

6. Finally, we will style all of the odd-numbered list items after the selected item using the all-next-siblings combinator. In this case, the list items numbered 7 and 9 will have an orange background because they are the odd-numbered list items that are also siblings after the selected item (the list item numbered 5):

```
li.selected ~ li:nth-child(odd) {
    background-color: orange;
}
```

The result is seen in the following figure:

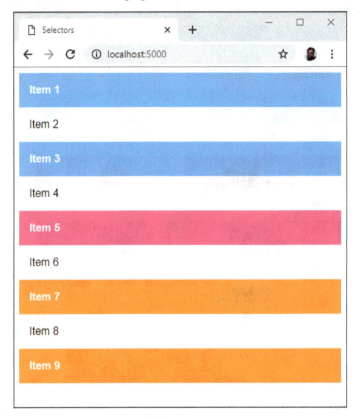

Figure 1.22: Combining selectors to style a list

Style sheets can have a large number of style rules and combinations of selectors. It is good to understand why one rule takes precedence over another one. This is where CSS specificity comes in.

CSS specificity

If we have two CSS declarations that affect the same style property of an element, how do we know which of those declarations will take precedence?

Several factors decide the ranking of a CSS declaration and whether it is the style the browser will apply. The term for these factors is **specificity**.

A style attribute that adds inline styles to an element has the highest specificity value. An ID selector has a greater specificity value than a class selector and a class selector or attribute selector has a greater specificity value than an element type. We can calculate the specificity value by giving points to each of these specificity values.

The most common way of representing this is as a comma-separated list of integers, where the leftmost integer represents the highest specificity. In other words, the leftmost value is the inline style attribute; next is an ID selector; next is a class selector, pseudo-class, or attribute selector; and the rightmost value is an element.

An inline style would have the 1, 0, 0, 0 value. An ID selector would have the 0, 1, 0, 0 value. A class selector would have the 0, 0, 1, 0 value, and an h1 selector would have the 0, 0, 0, 1 value.

Let's look at a few examples with more complex selectors:

- li.selected a[href] has two element selectors (li and a), a class selector (.selected), and an attribute selector ([href]), so its specificity value would be 0, 0, 2, 2:

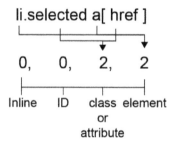

Figure 1.23: Calculating the specificity of li.selected a[href]

- #newItem #mainHeading span.smallPrint has two ID selectors, a class selector (.smallPrint), and a span element, so its specificity value would be 0, 2, 1, 1:

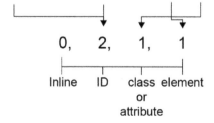

Figure 1.24: Calculating the specificity of #newItem #mainHeading span.smallPrint

Comparing the two selectors, we can see that the selector in the second example is more specific than the selector in the first example.

The special case of !important

The !important keyword can be appended to the value of any CSS declaration. It sets the specificity of that rule to have a special value of 1, 0, 0, 0, 0, which will give it precedence over any style including inline styles.

As an example of where it can be useful, we might want to create a style rule that is reusable and lets us hide content on a web page. If we apply this class to an element, we want that element to be hidden and not be rendered on the web page. However, consider the following example:

```
<style>
div.media {
  display: block;
  width: 100%;
  float: left;
}
.hide {
  display: none;
}
</style>
<div class="media hide">
  ...Some content
</div>
```

We might expect our div element to be hidden because the .hide class appears second in the style sheet. However, if we apply the specificity calculations we've learned about, we can see that div. media scores 0, 0, 1, 1, and .hide only scores 0, 0, 1, 0. The div.media rule for the display property with a block value will override the none value of the .hide class. We can't really use this instance of the .hide class as we don't know whether it will have any effect.

Now, consider the same .hide class but using the !important keyword:

```
.hide {
  display: none !important;
}
```

Adding the !important keyword will make this .hide class much more reusable and useful as we can pretty much guarantee that it will hide content as we desire.

> **Note**
>
> Using !important should always be a last resort as it takes the rule out of the usual CSS specificity and makes overriding styles more difficult.

CSS custom properties

A relatively new addition to CSS, CSS custom properties (often called **CSS variables**) allow you to store a value with a name and reuse that value in multiple different CSS rules.

A CSS variable is defined as a name and a value inside a CSS ruleset. The name is prefixed with a double hyphen, --. The value can be any valid CSS value. The selector part of the ruleset will specify the scope of the variable. The :root selector is often used to define variables for the whole document but the scope could be targeted to a specific part of the document.

When applying a CSS variable we use the var() function. An optional fallback value can be given for cases where the variable has not been set.

For example, we could set a --color-primary variable with the color hex value of #FC9C9C on the :root of the document and then later access that variable to set the color of all paragraphs in the document:

```
:root {
  --color-primary: #FC9C9C;
}

p {
  color: var(--color-primary, #FF0000);
}
```

The benefit of storing a value as a variable is that you can provide a semantic name for a variable and have that used in many different places keeping, for example, the colors for a theme consistent, more maintainable, and easier to update in the future.

We've learned a lot about the syntax and fundamentals of CSS in this chapter. Let's apply some of this knowledge to an activity.

Activity 1.02 – styling the video store template page

In the previous activity, we were tasked with creating boilerplate HTML for a web page for the *Films on Demand* website. In this activity, we are going to add some style to that template page.

The steps are as follows:

1. We will start with the template from *Activity 1.01*, which we will save as `template.html`:

```
<!DOCTYPE html>
<html lang="en">
    <head>
        <meta charset="utf-8">
        <title>
            Films on Demand -
            <!-- Title for page goes here -->
        </title>
        <meta name="description"
            content="Buy films from our great
                    selection. Watch movies on
                    demand.">
        <meta name="viewport"
            content="width=device-width,
                    initial-scale=1">
    </head>
    <body>
        <h1>Lorem ipsum</h1>
        <p>
            Lorem ipsum dolor sit amet, consectetur
            adipiscing elit. Nullam quis scelerisque
            mauris. Curabitur aliquam ligula in erat
            placerat finibus. Mauris leo neque,
            malesuada et augue at, consectetur rhoncus
            libero. Suspendisse vitae dictum dolor.
            Vestibulum hendrerit iaculis ipsum, ac
            ornare ligula. Vestibulum efficitur mattis
            urna vitae ultrices. Nunc condimentum
            blandit tellus ut mattis. Morbi eget
            gravida leo. Mauris ornare lorem a mattis
            ultricies. Nullam convallis tincidunt
            nunc, eget rhoncus nulla tincidunt sed.
            Nulla consequat tellus lectus, in porta
            nulla facilisis eu. Donec bibendum nisi
            felis, sit amet cursus nisl suscipit ut.
            Pellentesque bibendum id libero at cursus.
            Donec ac viverra tellus. Proin sed dolor
            quis justo convallis auctor sit amet nec
            orci. Orci varius natoque penatibus et
```

```
                    magnis dis parturient montes, nascetur
                    ridiculus mus.
                </p>
            </body>
        </html>
```

2. We are going to link to an external CSS file. One of the difficulties with styling web pages is handling differences between browsers. We are going to do this by adding a file to normalize our default styles. We will use the open source `normalize.css` for this. Download the file from `https://packt.link/kBTXT`. Add the file to a styles folder and link to it from the `template.html` web page.

3. We are going to add a `style` element to the `head` element of `template.html`. In the `style` element, we want to set some styles used across all pages. We want to do the following:

    • We want to set `box-sizing` to `border-box` for all elements using the universal selector (`*`).

    • We want to add a font family with the `Arial`, `Helvetica`, and `sans-serif` values and a font size of `16` px to the whole page.

    • We want to add the `#eeeae4` background color for the whole page. To do this, we will add a `div` element wrapper with the `pageWrapper` ID, where we will set the background color and padding to `16` px, and a `full-page` class, where we will set the minimum height to `100` vh (100% of the viewport height).

    • We want to add an `h1` element selector that sets `margin` to `0` and adds `padding` of `16` px to the bottom of the `h1` element.

> **Note**
>
> The solution to this activity is available on GitHub at `https://packt.link/WbEPx`.

In this section, we have used CSS to add color and style to a web page. We have looked at the syntax of CSS and learned how we can use CSS selectors to apply style rules to specific parts of a web page. We have also learned a way to calculate the specificity of CSS rules and we have learned about CSS custom properties and how these can help make code more maintainable and understandable.

Summary

In this chapter, we have looked at how we write HTML and CSS and their roles in how a browser understands and renders a web page.

To demonstrate our understanding, we have created a web page that we will use as a template for the pages of a website, and we have added CSS to normalize the style and add some initial styles for the site. Using the skills we have learned, we can now create a simple web page, add metadata to it, and style parts of that web page as we choose.

In the next chapter, we will look at the options for structuring and laying out a web page, the elements within it, and how we can style these for different web page layouts.

2

Structure and Layout

In the previous chapter, we learned about the basics of **HTML** and **CSS**. In this chapter, we will consolidate this basic understanding and look at how web pages are structured with HTML and CSS. When creating web pages using HTML, it is imperative that you use the correct elements. This is because HTML is read by both humans and machines and so the content of a web page should be associated with the most appropriate element. Additionally, any error in the code might be difficult to track if the code base is too large.

The HTML language offers a vast array of different tags that we can place at our disposal. In this chapter, we will focus on the structural elements that are used to divide the web page into its key parts. You may be familiar with the concept of a page header or footer, and these would be examples of structural elements. We will be looking at these amongst many other HTML structural elements.

In this chapter, we will focus our attention on the **HTML5** version of the language, which is the most current version of HTML. HTML5 offers us additional tags that enable us to make our markup more meaningful. The developer experience is more enjoyable compared to writing **XHTML** as the HTML5 language is less strict with regard to syntax.

> **Note**
> XHTML is a term used to describe HTML written in conformance with XML syntax rules. Also, in this chapter, we will use the terms "tags" and "elements" synonymously.

Web pages are typically styled using CSS. Once we have our web pages marked up correctly, we need to know how to style these into a range of layouts. CSS offers us a range of options for laying out our pages, but the three most common methods are `float-`, `flex-`, and `grid`-based layouts. In this chapter, we will explore each of these techniques in turn.

Just knowing the various layout methods is not enough to style web pages. We will investigate the box model, which is foundational to understanding how HTML elements are styled. We will break this down into individual layers – the `content` box, `padding`, `border`, and `margin`. With this knowledge in hand, you will be free to develop a host of different web page layouts.

By the end of this chapter, you will be able to use the correct HTML5 elements to mark up a web page; style a web page using float, flex, and grid layouts; describe how the box model works; and build a home page and a product page layout.

The following topics will be covered in the chapter:

- Using correct HTML5 elements to markup a web page
- Styling a web page using `float`-, `flex`-, and `grid`-based layouts
- Describing how the box model works

We will now take a look at the structural elements provided by HTML and examine what the key elements are one by one.

Technical requirements

The code files for this chapter can be found at `https://packt.link/c49Po`.

Examining structural elements

HTML5 provides us with a variety of tags that we can use when dividing our page into different parts. When browsing the web, you would have noticed that web pages typically have a few common things to them. For example, a web page will typically have a logo and page navigation area at the top of the page. We would call this area of the page the **header**. You may also have noticed that the bottom of the page may include a list of links and copyright information. We would call this area the **footer**. The following diagram shows the representation of a few of the main elements of a web page:

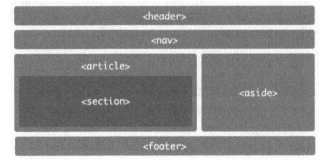

Figure 2.1: HTML5 page elements

In this section, we will be looking at the following HTML5 page elements:

- header
- footer

- `section`
- `article`
- `nav`
- `aside`
- `div`

header

The `header` tag is used to describe the header or top area of a web page. Typically, inside this tag, you would have the page heading, a logo, and, possibly, the page navigation. Prior to HTML5, you would use a `div` tag with a class name so that the header could be styled, and its intention was clear to developers. HTML5 improves on this by giving us a tag specifically for this very task. You will learn more about this improvement under the *Writing semantic markup* section in *Chapter 3*. Now, examine the following code that shows the difference between the old and new ways of writing the markup for the `header` area:

```
<!-- old way -->
<div class="header">
  … heading, logo, nav goes here
</div>
<!-- new way -->
<header>
  … heading, logo, nav goes here
<header>
```

Now, let's open the Packt website at `https://www.packtpub.com/` to see how a header is represented in an actual website. In the following diagram, you can see that the `header` element is highlighted in red, illustrating where a `header` element is typically placed on a web page:

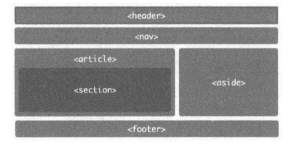

Figure 2.2: The header element

In the following figure, you can see that the header element is highlighted in red. As this is an example taken from the Packt website, you will notice that it contains items such as the company logo, search bar, and the **Sign In** button:

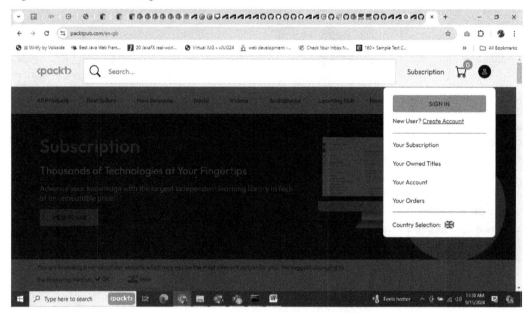

Figure 2.3: The header element on the Packt site

footer

The footer tag is very similar to the header tag but is used at the bottom of a web page. You would typically have the copyright information and website links inside the footer. Similarly, with the header tag in the previous version of HTML, you would use a div tag with a class name. Since the use of footers on web pages is so common, HTML5 provides a new tag solely for this purpose. The following code shows the difference between the old and new way of writing the markup for the footer area:

```
<!-- old way -->
<div class="footer">
  … copyright, list of links go here
</div>
<!— new way -->
<footer>
  … copyright, list of links go here
<footer>
```

In the following figure, you can see that the `footer` element is highlighted in red, illustrating where a `footer` element is typically placed on a web page:

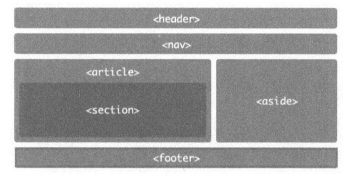

Figure 2.4: The footer element

In the following figure, you can see that the `footer` element is highlighted in red. As this is an example taken from the Packt website, you will notice it contains items such as **Useful Links** and social media icons:

Figure 2.5: The footer element on the Packt site

section

The `section` tag is different from the `header` and `footer` tags as it can be used in many different places on a web page. Some examples of when you would use a `section` tag could be for the main content area of a page or to group a list of related images together. You use this tag anytime you want to divide some of the markup into a logical section of the page. Again, prior to HTML5, you would most likely use a `div` tag with a class name to divide a section of the page. The following code shows the difference between the old and new way of writing the markup for the `section` area:

```
<!-- old way -->
<div class="main-content-section">
  ... main content
```

```
</div>

<!— new way -->
<section>
  ... main content
</section>
```

In the following figure, you can see that the `section` element is highlighted in red, illustrating where a section element is typically placed on a web page:

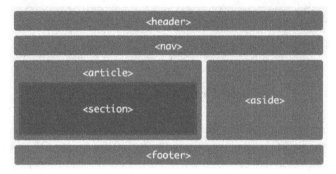

Figure 2.6: The section element

article

The `article` tag is used for the self-contained part of a web page. Some examples of an article could be an individual news article or blog post. You can have multiple articles on a page, but each must be self-contained and not dependent on any other context within the page. It is common to see the `article` tag used in conjunction with `section` tags to divide up an article into discrete sections. The following code shows this:

```
<article>
    <section>
      ...primary blog content
    </section>
    <section>
      ...secondary blog content
    </section>
</article>
```

In the following figure, you can see that the article element is highlighted in red, illustrating where an article element is typically placed on a web page:

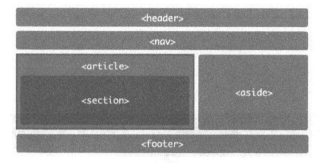

Figure 2.7: The article element

nav

Inside the navigation area, you will have a list of page links for the different pages of the website. Prior to HTML5, you would again use a `div` tag with a class name. The following code shows the difference between the old and new way of writing the markup for the navigation area:

```
<!-- old way -->
<div class="navigation">
   … list of links go here
</div>

<!-- new way -->
<nav>
   … list of links go here
</nav>
```

In the following figure, you can see that the `nav` element is highlighted in red, illustrating where a `nav` element is typically placed on a web page:

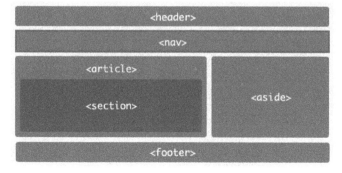

Figure 2.8: The nav element

In the following figure, you can see that the nav element is highlighted in red. As this is an example taken from the Packt website, you will notice it contains a list of page links:

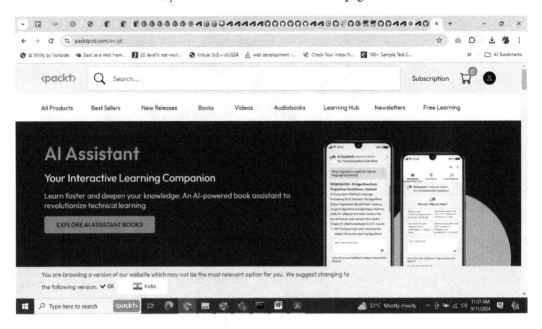

Figure 2.9: The nav element on the Packt site

aside

The aside tag is used to show content that is indirectly related to the main content of a document. You will typically see this tag used for sidebars or for showing notes relating to some content. Again, before the advent of HTML5, developers would use a div tag with a class name for this type of content. The following code shows the difference between the old and new way of writing the markup for the aside element:

```
<!-- old way -->
<div class="sidebar">
  … indirectly related content goes here
</div>

<!—new way -->
<aside>
  … indirectly related content goes here
</aside>
```

In the following figure, you can see that the `aside` element is highlighted in red, illustrating where an aside element is typically placed on a web page:

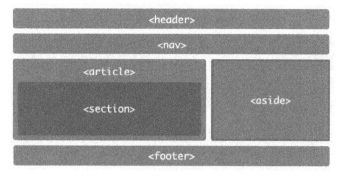

Figure 2.10: The aside element

div

The `div` tag is probably the most widely used tag on the World Wide Web. In fact, if you view the source code of your favorite website, most of the HTML elements you see will be `div` elements. This tag actually stands for *division* and is used to divide or group content together. Although HTML5 provides specialist elements for the most common types of page groups, you will still find many uses for using `div` tags. It might help to think of this element as a generic way to group the markup into logical parts. The following are a few pieces of example code of how a `div` tag may be used:

```
<div class="sidebar">
  … indirectly related content goes here
</div>

<div class="navigation">
  <div class="navigation-inner">... navigation links go
    here
  </div>
</div>
```

That concludes our tour of the structural HTML elements that are important to us. We will now apply some of this theory with the help of an exercise.

A news article web page

Now that we have an understanding of the structural elements provided by HTML5, let's put our newly acquired knowledge into practice by writing the structural HTML for a news article page. You can get a sense of what this type of page will look like by visiting a popular online news website such as https://theguardian.com or https://bbc.co.uk/news and clicking on an article.

Exercise 2.01 – marking up the page

In this exercise, we will create the markup for our HTML5 page. Our aim will be to produce a page with output, as shown in the following figure:

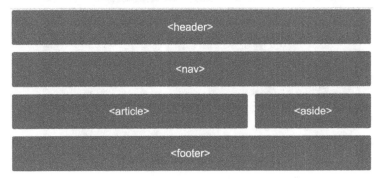

Figure 2.11: Expected output for the product page

> **Note**
> The complete code for this exercise can be found at https://packt.link/F3YGx.

Let's complete the exercise with the following steps:

1. Create a file named news.html in VSCode.

2. We will use the following starter HTML document, which contains some basic styling for our structural elements. Don't worry if you don't understand the CSS just yet; you will by the end of this course:

```
<!DOCTYPE html>
<html>
    <head>
        <title>News article page</title>
        <style>
        :root {
            --bg-color: #659494;
            --text-color: #fff;
            --size-large: 30px;
            --size-medium: 20px;
        }
```

```
header,
nav,
article,
aside,
footer {
  background: var(--bg-color);
  border-radius: 5px;
  color: var(--text-color);
  font-family: arial, san-serif;
  font-size: var(--size-large);
  text-align: center;
  padding: var(--size-large);
  margin-bottom: var(--size-medium);
}


header:before,
nav:before,
article:before,
aside:before,
footer:before {
  content: '<';
}


header:after,
nav:after,
article:after,
aside:after,
footer:after {
  content: '>';
}


article {
  float: left;
  margin-right: var(--size-medium);
  width: 60%;
}
```

```
        aside {
            float: left;
            width: calc(40% - 140px);
        }


        footer {
            clear: both;
        }

    </style>
</head>
<body>
  <!-- your code will go here -->
</body>
</html>
```

3. First, let's add our first structural element, which is the `header` tag. We will place it in between the opening and closing `body` tags. In this example, we will just add some text as content but, when building a real web page, you would include things such as logos, search bars, and links:

```
<body>
    <header>header</header>
</body>
```

4. After our `header` tag comes the navigation area, which is used for including links to different pages of the website. Once again, we will just add some text for the content but, when building a real web page, you would include a list of links:

```
<body>
    <header>header</header>
    <nav>nav</nav>
</body>
```

5. For the main news article content, we will use an `article` tag. Once again, we will just add some text for the content but, when building a real web page, you would include the content of the articles:

```
<body>
    <header>header</header>
    <nav>nav</nav>
    <article>article</article>
</body>
```

6. To the right of the `article` tag, we have an `aside` tag, which will typically contain content such as advertising images or related content links:

```
<body>
    <header>header</header>
    <nav>nav</nav>
    <article>article</article>
    <aside>aside</aside>
</body>
```

7. Finally, we can finish off the markup for our web page by adding the `footer` tag at the bottom of the page. For now, we will just add some text as content but, in real life, you would include elements such as copyright information and links to other pages:

```
<body>
    <header>header</header>
    <nav>nav</nav>
    <article>article</article>
    <aside>aside</aside>
    <footer>footer</footer>
</body>
```

If you now right-click on the filename in VSCode on the left-hand side of the screen and select **Open in default browser**, you will see the web page in your browser.

If you look at this page in your browser, you may not be impressed with what you see, but you actually have the foundations in place for a web page.

In this section, we learned the most common HTML structural elements including `header`, `nav`, and `div` elements. We got to practice using these elements by creating pages for our news article web page.

In the next section, we will briefly learn about wireframes, which are used heavily in commercial web design project teams. We will again get some more practice with our newly learned HTML elements and build a video store home page.

Introducing wireframes

When working on commercial projects, it is common for web page designs to be provided to web developers in the form of a **wireframe**. A wireframe is a low-fidelity design that provides enough information about a page for the developer to start coding. Usually, they will not include much visual design information and are focused on the main structure of a page.

The following figure is an example of a wireframe for a new home page:

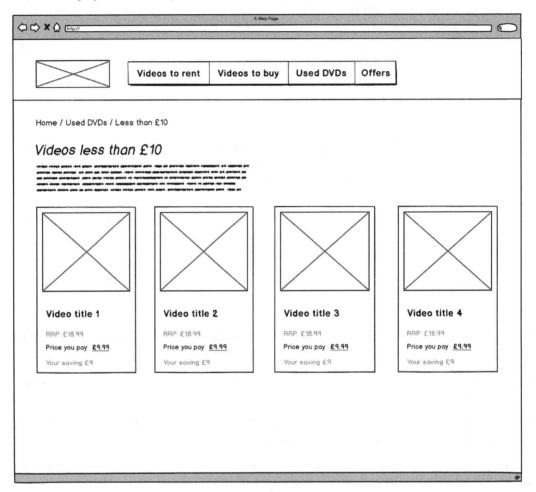

Figure 2.12: Example of a wireframe

Activity 2.01 – video store home page

Suppose you are a frontend developer working for a tech start-up. You have been asked to build a home page for the online video store. You have been given the following wireframe from the UX designer:

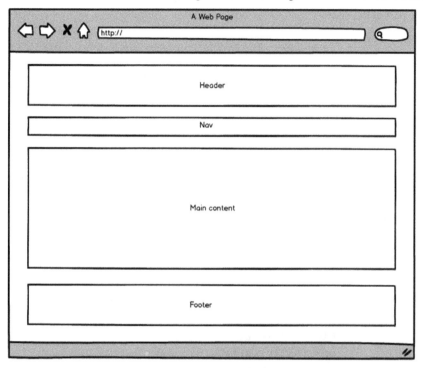

Figure 2.13: Wireframe as per the UX designer's expectation

Using your newly acquired HTML5 knowledge, you can start to convert the wireframe into working HTML code. At this stage, you should just be concerned with writing the structural HTML tags and shouldn't worry about content right now. The aim will be to achieve a web page like the following output screenshot:

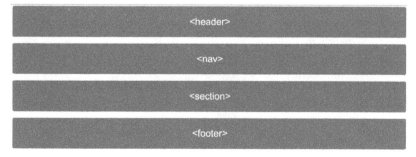

Figure 2.14: Expected output of video store home page

The steps are as follows:

1. Create a file named home.html in VSCode.

2. Use the following code as a page skeleton. Again, do not worry about not understanding the styling part of the code:

```html
<!DOCTYPE html>
<html>
    <head>
        <title>Video store home page</title>
        <style>
          :root {
            --bg-color: #659494;
            --text-color: #fff;
            --size-large: 30px;
            --size-medium: 20px;
          }

          header,
          nav,
          section,
          footer {
            background: var(--bg-color);
            border-radius: 5px;
            color: var(--text-color);
            font-family: arial, san-serif;
            font-size: var(--size-large);
            text-align: center;
            padding: var(--size-large);
            margin-bottom: var(--size-medium);
          }

          header:before,
          nav:before,
          section:before,
          footer:before {
            content: '<';
          }

          header:after,
          nav:after,
          section:after,
```

```
           footer:after {
             content: '>';
           }

       </style>
     </head>
     <body>
   <!-- your code will go here -->
     </body>
   </html>
```

3. Start adding the HTML5 structural elements inside the body tag one by one, the same as we did in *Exercise 2.01, marking up the page*.

4. As with *Exercise 2.01, marking up the page*, we will just add the tag name for content such as header and footer.

If you now right-click on the filename in VSCode on the left-hand side of the screen and select **Open in default browser**, you will see the web page in your browser.

Hopefully, you are now getting a feel for the process of putting basic web pages together. We will build on this knowledge in the coming exercises.

> **Note**
>
> The solution to this activity can be on GitHub at https://packt.link/rMCWY.

We are now ready to start making our web pages more realistic by learning some CSS page layout techniques.

Learning CSS page layouts

CSS provides us with a range of possibilities for laying out web pages. We will be looking into the three most common techniques for laying out web pages. These are as follows:

- float
- flex
- grid

Armed with this knowledge, combined with your knowledge of HTML structural tags, you will be able to code a range of web page layouts. The concepts learned in this part of the chapter will form the core of your frontend development skillset and you will use these techniques over and over throughout your career.

Video store product page

In order to gain a solid understanding of how these three different approaches to layout work, we shall use a video store product listing page as a concrete example. We will work through solutions to the following design using the three most common layout techniques, one by one. For the examples that follow, we will only be concerned with the product section of the page:

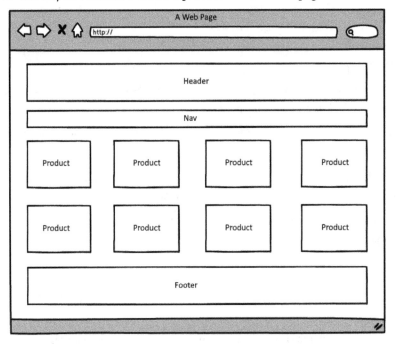

Figure 2.15: Product page wireframe

float-based layouts

The `float`-based CSS layout technique is the oldest of the three. Whilst CSS provides us with improved techniques for layout, the `float`-based layout is still used today. Having a firm grasp of how `float`-based layouts work in practice will set you up for more advanced styling segments in this book.

float property

The CSS `float` property, when applied to an element, will place the element to either the left or right of its containing element. Let's examine a few examples of the most common values for this property.

To float elements to the right, you would use the `right` value, as shown in the following code:

```
float: right;
```

Whereas to float elements to the left, you would use the `left` value, as shown in the following code:

```
float: left;
```

The `none` value isn't used as frequently but, with the following code, it can be handy if you wish to override either the left or right values:

```
float: none;
```

width property

When we apply the `float` property to elements, we typically will also want to give the element an explicit `width` value as well. We can either give a value in pixels or percentages. The following code shows the input for `width` in pixels, that is, by writing `px` after the value:

```
width: 100px;
```

The following code shows the input for `width` as a percentage, that is, by entering the `%` symbol after the value:

```
width: 25%;
```

Clearing floated elements

As the name suggests, floated elements do, in fact, appear to float in relation to the other non-floated elements on the page. A common issue with floated elements inside a container is illustrated in the following figure:

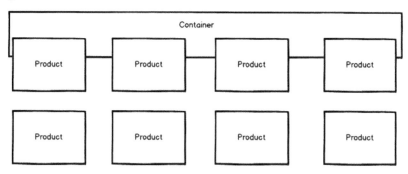

Figure 2.16: Floating elements illustration

Note
This solution to clearing floated elements has been used for simplicity.

There are many solutions to this issue, but by far the easiest solution is to apply the following CSS to the containing element:

```
section {
    overflow: hidden;
}
```

With the preceding code added to the container, we will now have floated elements contained inside the wrapping element, as illustrated in the following figure:

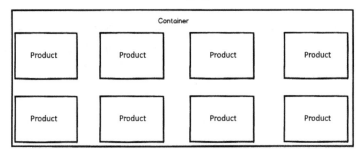

Figure 2.17: Cleared floats illustration

The following example code shows how you could achieve the preceding layout using `float`:

```
<!-- HTML -->
<section>
    <div>product 1</div>
    <div>product 2</div>
    <div>product 3</div>
    <div>product 4</div>
    <div>product 5</div>
    <div>product 6</div>
    <div>product 7</div>
    <div>product 8</div>
</section>

/* CSS */
section {
    overflow: hidden;
}

div {
    float: left;
    width: 25%;
}
```

To take a deeper dive into more advanced uses of the `float` property, you can have a look at the **Mozilla Developer Network** MDN page here: `https://developer.mozilla.org/en-US/docs/Web/CSS/float`.

flex-based layouts

The `flex`-based CSS layout technique is a new and improved alternative to the `float`-based approach. With `flex`, we have much more flexibility and can easily achieve complex layouts with very little code. With `flex`, we no longer have to worry about clearing floating elements. We will now look into some of the key properties and values in order to let us build the product page layout using `flex`.

flex container

When developing `flex`-based layouts, there are two key concepts you must first understand. The first is the `flex` container, which is the element that contains the child elements. To activate a `flex` layout, we must first apply the following code to the container or parent element that holds the individual items:

```
display: flex;
```

We also have to choose how we want the container to handle the layout of the child elements. By default, all child elements will fit into one row. If we want the child elements to show on multiple rows, then we need to add the following code:

```
flex-wrap: wrap;
```

flex items

Now that we know how to set the `flex` container up, we can turn to the child elements. The main issue of concern here is the need to specify the width of the child elements. To specify this, we need to add the following code:

```
flex-basis: 25%;
```

You can think of this as being equivalent to the width in our `float`-based example.

The following example code shows how you could achieve the product layout, as shown in *Figure 2.15*, using `flex`:

```
<!-- HTML -->
<section>
  <div>product 1</div>
  <div>product 2</div>
  <div>product 3</div>
  <div>product 4</div>
  <div>product 5</div>
```

```
    <div>product 6</div>
    <div>product 7</div>
    <div>product 8</div>
</section>

/* CSS */
section {
  display: flex;
  flex-wrap: wrap;
}

div {
  flex-basis: 25%;
}
```

To take a deeper dive into more advanced uses of the `flex` property, you can have a look at the MDN page here: `https://developer.mozilla.org/en-US/docs/Web/CSS/flex`.

grid-based layouts

The `grid`-based CSS layout technique is the newest of the three different approaches we will be exploring. This new approach was introduced in order to simplify the page layout and offer developers even more flexibility vis-à-vis the previous two techniques. We will now look into some of the key properties and values to enable us to build the product page layout using a `grid`-based approach.

grid container

When developing `grid`-based layouts, there are two key concepts you must first understand. The first is the grid container, which is the element that contains the child elements. To activate a grid layout, we must first apply the following code to the parent element:

```
display: grid;
```

Now that we have activated the container to use the `grid`-based layout, we need to specify the number and sizes of our columns in the grid. The following code would be used to have four equally spaced columns:

```
grid-template-columns: auto auto;
```

grid items

When we used `float` and `flex` layouts, we had to explicitly set the width of the child elements. With `grid`-based layouts, we no longer need to do this, at least for simple layouts.

We will now put our new-found knowledge into practice and build the product cards shown in *Figure 2.15*. We will use the grid layout technique since the product cards are actually within a `grid` layout, comprising four equally spaced columns.

To take a deeper dive into more advanced uses of the `grid` property, you can have a look at the MDN page here: `https://developer.mozilla.org/en-US/docs/Web/CSS/grid`.

Exercise 2.02 – a grid-based layout

In this exercise, we will create our CSS page layout with the aim of producing a web page where six products are displayed. The following figure shows how the output should appear:

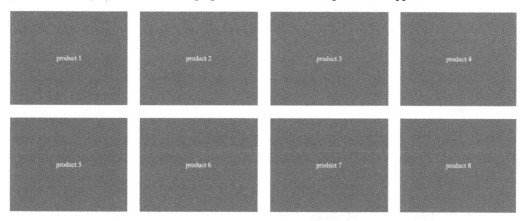

Figure 2.18: Expected output for the grid-based layout

> **Note**
>
> The complete code for this exercise can be found at `https://packt.link/H3VBm`.

The following are the steps to complete this exercise:

1. Let's begin with the following HTML skeleton and create a file called `grid.html` in VSCode. Don't worry if you do not understand the CSS used here; you will soon enough:

    ```
    <!DOCTYPE html>
    <html>
    <head>
      <title>Grid based layout</title>
      <style type="text/css">
        :root {
    ```

```
            --bg-color: #659494;
            --text-color: #fff;
            --size-large: 100px;
            --size-small: 20px;
    }

    div {
        background: var(--bg-color);
        color: var(--text-color);
        text-align: center;
        margin: var(--size-small);
        padding: var(--size-large);
    }
    </style>
</head>
<body>
</body>
</html>
```

2. Next, we will add the product items using `div` tags, which are placed inside a `section` tag. We will just add a product with a number inside each item, so we know what product each represents:

```
<body>
  <section>
    <div>product 1</div>
    <div>product 2</div>
    <div>product 3</div>
    <div>product 4</div>
    <div>product 5</div>
    <div>product 6</div>
    <div>product 7</div>
    <div>product 8</div>
  </section>
</body>
```

3. Now, let's add the following CSS in order to activate the `grid`-based layout. If you compare this to the other two techniques for laying out web pages, the code is very minimal:

```
section {
  display: grid;
  grid-template-columns: auto auto auto auto;
}
```

If you now right-click on the filename in VSCode on the left-hand side of the screen and select **Open in default browser**, you will see the web page in your browser.

If you now look at this page in your web browser, you should see a layout resembling the one shown in the screenshot.

We will now take a detour and look into some fundamental concepts of how CSS styles HTML elements.

The box model

So far, all the elements on our pages look almost identical because we have not learned how to adjust the size of each element. We are now ready to progress to more realistic page designs by introducing a foundational layout concept called the box model.

Try to picture each HTML element as a box made up of different layers. The different layers are the element's content box, padding, border, and margin. We will explore each of these layers one by one. The following figure illustrates how all aspects of the box model relate to one another. You can see that the margin is the outermost part, followed by the element's border and padding between the border and content area:

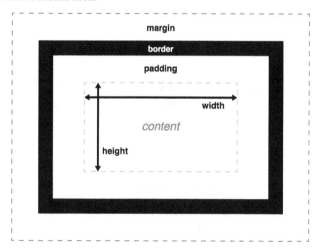

Figure 2.19: The box model

We will now look at each of the box model elements, in turn, starting with the innermost content box.

The content box

The content box is the part of the element where the actual content lives. This is typically text but could contain other child elements or media elements such as images. The most important CSS properties for this layer are width and height. As a developer, you would typically give these values expressed in pixels or percentages. The following code shows some example values, followed by the corresponding output figure for these properties:

```
width: 200px;
height: 100px;
```

In the following figure, we will see what the content area looks like after CSS is applied to the preceding code:

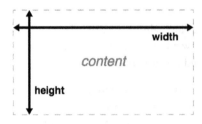

Figure 2.20: The content box

Next, we will work our way out to the next layer of the box model – padding.

padding

The padding area is the layer that provides spacing between the content box and the border. The amount of spacing in this layer can be specified in all directions – top, right, bottom, and left. CSS provides a padding property where you can specify values for the amount of spacing in all directions. If you want to apply the same amount of padding in all directions, you can just give a single value. If you want to apply the same values for vertical and horizontal directions, you can specify two values. It also provides direction-specific properties – padding-top, padding-right, padding-bottom, and padding-left. The following code shows a number of example values for these properties:

```
/* 50px of padding applied in all directions */
padding: 50px;

/* 50px of padding applied vertically and 0px applied horizontally */
padding: 50px 0;

/* 10px of padding applied to the top */
padding-top: 10px;

/* 10px of padding applied to the right */
padding-right: 10px;
```

```
/* 10px of padding applied to the bottom */
padding-bottom: 10px;

/* 10px of padding applied to the left */
padding-left: 10px;
```

The following figure illustrates what the content and padding areas would look like after CSS is applied to the following code:

```
width: 200px;
height: 100px;
padding: 25px;
```

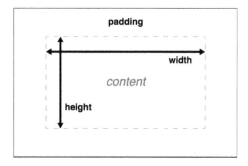

Figure 2.21: Padding

Now that we understand how the content and padding layers relate to one another, we will work our way out to the next layer of the box model – the border.

border

The border area is the layer that sits between the end of the padding area and the beginning of the margin. By default, the border isn't visible; it can only be seen when you explicitly set a value that will allow you to see the border. Similar to the padding property, CSS provides a shorthand property called border, and also the direction-specific properties – border-top, border-right, border-bottom, and border-left. All of these properties require three values to be provided: the width of the border, the border style, and finally, the color of the border. The following code shows some example values for these properties:

```
/* border styles applied in all directions */
border: 5px solid red;

/* border styles applied to the top */
border-top: 5px solid red;
```

```
/* border styles applied to the right */
border-right: 15px dotted green;

/* border styles applied to the bottom */
border-bottom: 10px dashed blue;

/* border styles applied to the left */
border-left: 10px double pink;
```

The following figure illustrates how the four different border styles would appear if applied to an element:

Figure 2.22: Border styles

The following figure illustrates what the `content`, `padding`, and `border` layers would look like after CSS is applied to the following code:

```
width: 200px;
height: 100px;
padding: 25px;
border: 10px solid black;
```

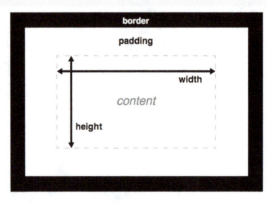

Figure 2.23: border

Now that we understand how the `content`, `padding`, and `margin` layers relate to one another, we will work our way out to the final layer of the box model – `margin`.

margin

The `margin` area is the layer that provides spacing between the edge of the border and out toward other elements on the page. The amount of spacing in this layer can be specified in all directions – top, right, bottom, and left. The CSS provides a `margin` property where you can specify values for the amount of spacing in all directions. It also provides direction-specific properties – `margin-top`, `margin-right`, `margin-bottom`, and `margin-left`. The following code shows a number of example values for these properties:

```
margin: 50px;
margin: 50px 0;

margin-top: 10px;
margin-right: 10px;
margin-bottom: 10px;
margin-left: 10px;
```

The following figure illustrates what the `content`, `padding`, `border`, and `margin` layers would look like after CSS is applied to the following code:

```
width: 200px;
height: 100px;
padding: 25px;
border: 10px solid black;
margin: 25px;
```

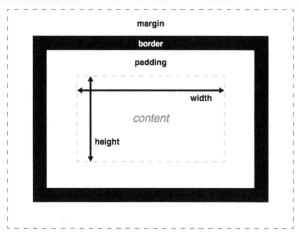

Figure 2.24: margin

To get some practice looking at how different HTML elements, make use of the box model; you can use the web tools inspector in your favorite browser. In Chrome, you can inspect an element and investigate how the box model is used for each element. If you inspect an element and then click the **Computed** tab on the right-hand side, you will see a detailed view. The following figure shows an example of an element from the Packt website revealing the values for properties from the box model:

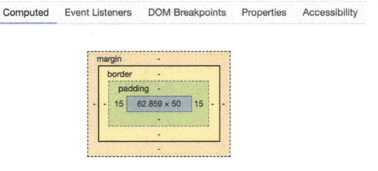

Figure 2.25: Chrome web tools box model inspection view

In the following exercise, we will play around with the different box model properties to get some practice with box model-related CSS properties.

Exercise 2.03 – experimenting with the box model

The aim of this exercise will be to create the three boxes as shown in the following output screenshot:

Figure 2.26: Expected boxes

> **Note**
> The complete code for this exercise can be found at https://packt.link/TwIuv.

The steps to complete the exercise are as follows:

1. First, let's add the following HTML skeleton to a file called `boxes.html` in VSCode:

```
<!DOCTYPE html>
<html>
<head>
  <title>Experimenting with the box model</title>
  <style type="text/css">
  </style>
</head>
<body>
  <div class="box-1">Box 1</div>
  <div class="box-2">Box 2</div>
  <div class="box-3">Box 3</div>
</body>
</html>
```

2. Now, let's add some CSS to the first box, observing the `width`, `height`, `padding`, and `border` properties we are adding. We will add the CSS in between the opening and closing style tags, as shown in the following code, to render the following figure:

```
<style type="text/css">
  .box-1 {
    float: left;
    width: 200px;
    height: 200px;
    padding: 50px;
    border: 1px solid red;
  }
</style>
```

Figure 2.27: Output for box 1

3. Now, let's add the CSS to the second box in *Figure 2.25*, observing how the `width`, `height`, `padding`, and `border` properties differ from the first box. We are using percentage-based measurements for the width and height properties, as shown in the following code, to render the following figure:

```css
.box-2 {
  float: left;
  width: 20%;
  height: 20%;
  padding-top: 50px;
  margin-left: 10px;
  border: 5px solid green;
}
```

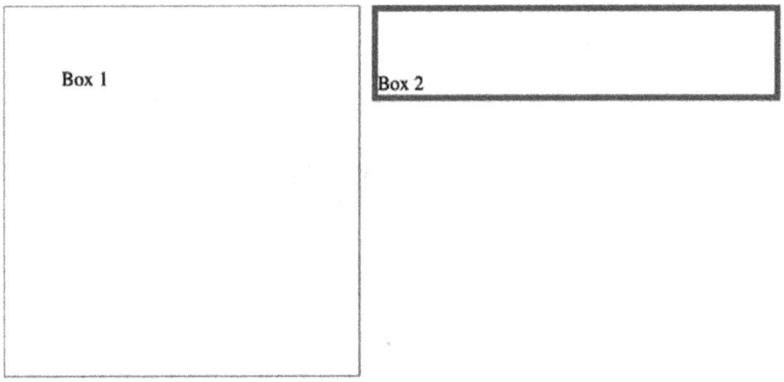

Figure 2.28: Output for boxes 1 and 2

4. Finally, let's add the CSS to the third box in *Figure 2.25*, observing how the `width`, `height`, `padding`, and `border` properties differ from the first and second boxes, as shown in the following code, to render the following figure:

```css
.box-3 {
  float: left;
  width: 300px;
  padding: 30px;
  margin: 50px;
  border-top: 50px solid blue;
}
```

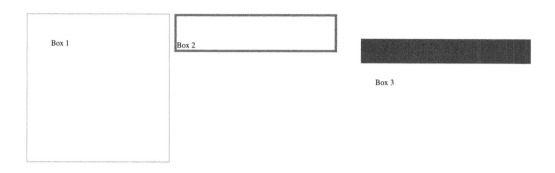

Figure 2.29: Output for boxes 1, 2, and 3

If you now right-click on the filename in VSCode on the left-hand side of the screen and select **Open in default browser**, you will see the web page in your browser.

This should give you a sense of what's possible with the box model. Feel free to change the various properties and experiment with different combinations.

Putting it all together

We now know how to correctly mark up a web page with the correct HTML5 structural tags. We also know how to use the three most popular CSS layout techniques. Finally, we have an understanding of how the box model works. We will now build the two complete web pages, combining all of the things we have learned so far in this chapter.

Exercise 2.04 – home page revisited

In this exercise, we will be using the wireframe in *Figure 2.13* for a home page design used in *Activity 2.01 – video store home page*. We will build a version of this page, incorporating the concepts from the box model topic. Our aim will be to build a page with the following output figure:

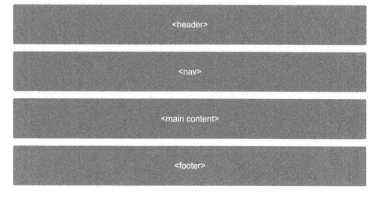

Figure 2.30: Expected output of home page

> **Note**
>
> The complete code for this exercise can be found at `https://packt.link/t9oEe`.

The steps to complete this exercise are as follows:

1. Create a new file called `home.html` in VSCode.

2. Use the following HTML code as a start file. Again, don't worry if some of the CSS doesn't make sense to you. We will look into this part of the styling in more detail in *Chapter 3*:

```html
<!DOCTYPE html>
<html>
    <head>
        <title>Video store home page</title>
        <style>
        :root {
            --bg-color: #659494;
            --text-color: #fff;
            }

        header,
        nav,
        section,
        footer {
            background: var(--bg-color);
            border-radius: 5px;
            color: var(--text-color);
            font-family: arial, san-serif;
            font-size: 30px;
            text-align: center;
        }

        header:before,
        nav:before,
        section:before,
        footer:before {
            content: '<';
        }
```

```
            header:after,
            nav:after,
            section:after,
            footer:after {
              content: '>';
            }
          </style>
      </head>
      <body>
        <header>Header</header>
      <nav>Nav</nav>
      <section>Main Content</section>
      <footer>Footer</footer>

      </body>
    </html>
```

3. Now, let's add some styling for the structural elements. Notice how we have used what we have learned from *The box model* topic to include `border`, `padding`, and `margin` with our structural elements. We will use a border to visually define the outer edge of the element, along with some padding to add spacing between the text and the outer edge of the element and a bottom `margin` to provide vertical spacing between the elements. We will add this just before the closing `style` tag:

```
/* CSS code above */
header,
nav,
section,
footer {
  border: 1px solid grey;
  padding: 50px;
  margin-bottom: 25px;
}
</style>
```

If you now right-click on the filename in VSCode on the left-hand side of the screen and select **Open in default browser**, you will see the web page in your browser.

You should now see a web page resembling the one shown in the home page wireframe.

Exercise 2.05 – video store product page revisited

In this exercise, we will be using the wireframe for a product page design as in *Figure 2.15*. We will build a more realistic version incorporating the box model. Our aim will be to build a page along the lines of the following output screenshot:

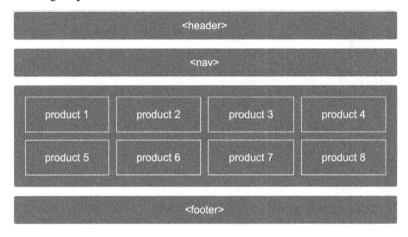

Figure 2.31: Expected output for the video store product page

> **Note**
> The complete code for this exercise can be found at https://packt.link/cLkov.

The steps to complete the exercise are as follows:

1. Create a new file called product.html in VSCode with the following code:

```
<!DOCTYPE html>
<html>
<head>
  <title>Video store product page</title>
  <style>
  </style>
</head>
<body>
</body>
</html>
```

2. In order to add styling, add the following code in between the `style` tags:

```css
:root {
        --bg-color: #659494;
        --text-color: #fff;
    }

header, nav, section, footer {
    background: var(--bg-color);
    border-radius: 5px;
    color: var(--text-color);
    font-family: arial, san-serif;
    font-size: 30px;
    text-align: center;
}
header:before, nav:before, footer:before {
    content: '<';
}
header:after, nav:after, footer:after {
    content: '>';
}
```

3. We will now add the HTML for the page elements, which are `header`, `nav`, `section`, and `footer`. The product items will be `div` elements inside the `section` element, as shown in the following code:

```html
<body>
    <header>header</header>
    <nav>nav</nav>
    <section>
        <div>product 1</div>
        <div>product 2</div>
        <div>product 3</div>
        <div>product 4</div>
        <div>product 5</div>
        <div>product 6</div>
        <div>product 7</div>
        <div>product 8</div>
    </section>
    <footer>footer</footer>
</body>
```

4. Now, let's add some styling for the structural elements. This is the same code as in the previous exercise. We will use a border to visually define the outer edge of the element, along with some padding to add spacing between the text and the outer edge of the element and a bottom margin to provide vertical spacing between elements. Again, we will add the CSS just before the closing `style` tag:

```css
/* CSS code above */

header,
nav,
section,
footer {
  border: 1px solid grey;
  padding: 20px;
  margin-bottom: 25px;
}
</style>
```

5. We will now need to add some styling for the product cards. We will use the `grid` layout technique, as this will allow our code to be as concise as possible:

```css
/* CSS code above */

section {
  display: grid;
  grid-template-columns: auto auto auto auto;
}

section div {
  border: 2px solid white;
  padding: 30px;
  margin: 10px;
}
</style>
```

If you now right-click on the filename in VSCode on the left-hand side of the screen and select **Open in default browser**, you will see the web page in your browser.

You should now see a web page resembling the one shown in the product page wireframe.

Activity 2.02 – online clothes store home page

Suppose you are a freelance web designer/developer and have just landed a new client. For your first project, the client wants a web home page developed for their online clothes store.

Using the skills learned in this chapter, design and develop the home page layout for the new online store.

The steps are as follows:

1. Produce a wireframe, either by hand or by using a graphics tool, for the new home page layout.

2. Create a file named `home.html` in VSCode.

3. Start writing out the markup for the page.

4. Now, style the layout with CSS.

The following figure shows the expected output for this activity:

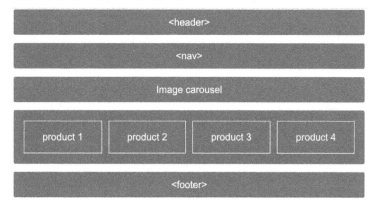

Figure 2.32: Expected output for the online clothes store home page

Note
The complete code for this activity can be found at `https://packt.link/kd3k5`.

Summary

In this chapter, we have begun our journey into building web pages. Knowing the range of HTML tags available to you is crucial in writing well-formed HTML documents. These include the `header`, `footer`, and `section` tags.

You should now feel comfortable taking a visual design or wireframe and converting this into the skeleton of an HTML document. We also looked at three common ways of styling a page layout with CSS. These involved the use of `float-`, `flex-`, and `grid`-based layout techniques. We then looked into what makes up the box model and used this knowledge to build the home and product pages of the video store.

In the next chapter, we will learn about the non-structural HTML elements used for content on a web page. We will then look into a number of common styling approaches to these elements using CSS.

3

Text and Typography Styling

In the previous chapter, you may have been frustrated that our web pages only dealt with the page structure and didn't contain any actual content. In this chapter, we will look at the HTML elements that are used for text-based page content. You are probably familiar with most of these elements from using word processing applications when writing documents. These HTML elements include elements such as headings, paragraphs, and lists.

As you are aware, when building web pages, the markup is only one aspect. We also need to style the elements of a page. We will take a look at issues concerning the styling of these text-based elements. We will also look into concerns such as cross-browser consistency. Finally, we will walk through some exercises to develop common web page components, such as navigation bars and breadcrumbs.

Text and typography are very important, as a visually appealing web page will ensure that a web user spends more time on it (which may be good for business conversion).

A highly functional, fast, and efficient website may not attract users if the text/typography is of poor quality. Thus, these are crucial elements while designing a web page, and this chapter equips you with the necessary tools in HTML/CSS to create visually appealing and aesthetic web pages.

The following topics will be covered in the chapter:

- Learn common text-based HTML elements
- Learn common styling techniques for these text-based elements
- Learn how to write high-quality semantic markup

By the end of this chapter, you will be able to identify the most suitable heading element for a web page, use the most common text-based HTML elements, develop common web page features such as navigation and breadcrumbs, explain the importance of semantic markup, and convert a design into semantic markup.

This chapter introduces the text-based elements used for web page content. We will first take a tour through the text-based HTML elements. We will then look into the styling of these elements, and finally, put this into practice with some exercises related to building common web page components.

Technical requirements

The code files for this chapter can be found at `https://packt.link/kgVfh`.

Creating text-based elements

HTML provides us with a variety of elements that are used for text-based content. While browsing the web, you might have noticed that web pages typically have similar text-based content. Most web pages will contain a page heading. The content will typically comprise headings, paragraphs, and lists. HTML equips you with tools to format such elements within a web page.

In this section, we will be looking at the following HTML text-based elements:

- Headings
- Paragraphs
- Inline text elements
- Lists

Headings

Heading elements in HTML offer six levels of hierarchy, ranging from h1 to h6. Now, h1 is typically only used once on a page, as it is the topmost heading for the document as a whole. The following code snippet shows how all of these headings are used and what they look like in the browser by default:

```
<h1>Heading level 1</h1>
<h2>Heading level 2</h2>
<h3>Heading level 3</h3>
<h4>Heading level 4</h4>
<h5>Heading level 5</h5>
<h6>Heading level 6</h6>
```

A web page from the preceding code would appear as follows:

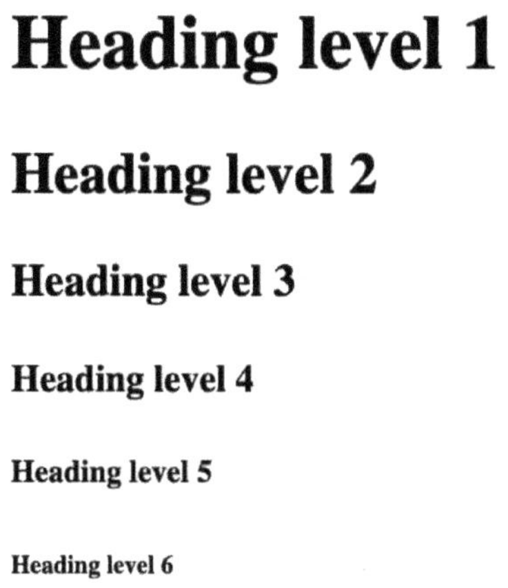

Heading level 1

Heading level 2

Heading level 3

Heading level 4

Heading level 5

Heading level 6

Figure 3.1: Headings shown in a browser

Paragraphs

Paragraphs in HTML can be represented using the p tag. On a web page, you might have chunks of the core content of a topic presented to the reader. Such content is included under a p tag in HTML. The following code snippet shows how you would include paragraphs in a document and what they look like by default in the browser:

```
<p>Horatio says 'tis but our fantasy,
And will not let belief take hold of him
Touching this dreaded sight, twice seen of us.
Therefore I have entreated him along,
With us to watch the minutes of this night,
That, if again this apparition come,
He may approve our eyes and speak to it.</p>

<p>Tush, tush, 'twill not appear.</p>

<p>Sit down awhile, And let us once again assail your ears,
That are so fortified against our story,
What we two nights have seen.</p>
```

> Note
>
> The text used for the preceding code is a work of Shakespeare from https://www.opensourceshakespeare.org/views/plays/play_view.php?WorkID=hamlet&Act=1&Scene=1&Scope=scene.

A web page for the preceding code will appear as follows:

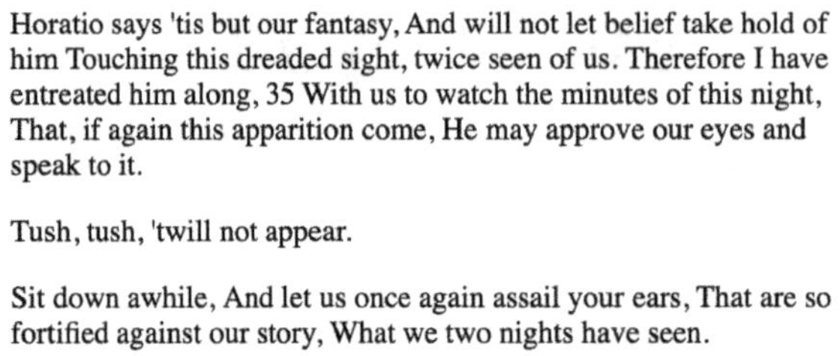

> Horatio says 'tis but our fantasy, And will not let belief take hold of him Touching this dreaded sight, twice seen of us. Therefore I have entreated him along, 35 With us to watch the minutes of this night, That, if again this apparition come, He may approve our eyes and speak to it.
>
> Tush, tush, 'twill not appear.
>
> Sit down awhile, And let us once again assail your ears, That are so fortified against our story, What we two nights have seen.

Figure 3.2: Paragraphs shown in a browser

Inline text elements

As a designer of web pages, you may often find yourself in a situation where you need to highlight special terms in a paragraph. Fortunately, HTML provides a solution to this.

It is possible to add what are called inline elements around text contained within paragraphs. Imagine using a word processor – you are able to make words bold, underlined, italicized, and so on. HTML provides developers with this ability, and we will now look at some of the most common examples.

- If you want to emphasize some text, you can use the em tag. An example of how you would use this and what it would look like in a browser is shown here:

```
<p>I need to wake up <em>now</em>!</p>
```

A web page for the preceding code would appear as follows:

I need to wake up *now*!

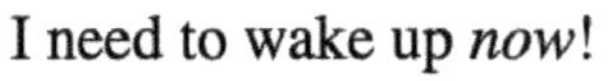

Figure 3.3: The em tag as it appears on a web page

- When you want to show some text that has serious importance, you can use the strong tag. An example of how you would use this and what it would look like in a browser is shown here:

```
<p>
  Before leaving the house
  <strong>remember to lock the front door</strong>!
</p>
```

A web page for the preceding code would appear as follows:

Before leaving the house **remember to lock the front door**!

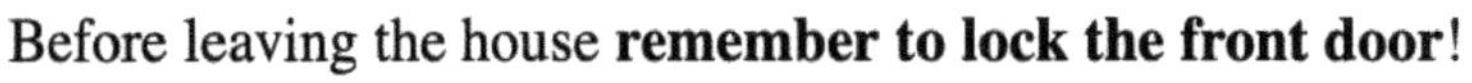

Figure 3.4: The strong tag as it appears on a web page

- Perhaps the most important of all the inline text-based elements is the anchor element, which allows you to add hyperlinks. To inline a link, you use an `a` tag wrapped around some text. An example of how you would use this and what it would look like in a browser is shown here:

```
<p>
   Please click
   <a href="http://www.google.com">here</a>
   to go to google
</p>
```

A web page for the preceding code would appear as follows:

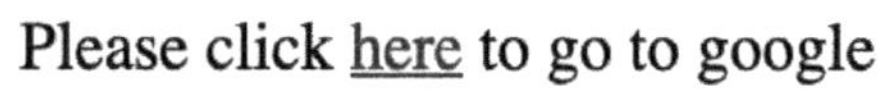

Please click <u>here</u> to go to google

Figure 3.5: Anchor as it appears on a web page

- Another important inline element for you to learn to use is the `span` tag. This is similar to the `div` tag but is used for inline elements. The `span` tag is used as a generic way to divide up content and has no inherent meaning, unlike the other inline tags mentioned previously in this chapter. A common use case is when styling a part of an element's content differently from the rest of the content. The following code shows an example of this:

```
/* styles */
.red {
   color: red;
}

.green {
   color: green;
}

.blue {
   color: blue;
}

<!-- markup -->
<p>
   My favorite colors are
   <span class="red">red</span>,
   <span class="green">green</span> and
   <span class="blue">blue</span>.
</p>
```

A web page for the preceding code would appear as follows:

My favorite colors are red, green and blue.

Figure 3.6: Paragraph with highlighted words as it appears on a web page

Lists

Another common type of text-based element that you will be very familiar with is the list. In HTML, these come in three different types – an unordered list, an ordered list, and a definition list. We will take a look at the differences between these types of lists and when you should use them.

1. Let's begin by taking a look at by far the most common type of list, the unordered list, which is expressed in HTML as ul, with li used for the list items. You will most likely be very familiar with this type of list in your everyday life. A common example of this type of list could be a shopping list or a list of things you need to pack before going on holiday. What makes this type of list unordered is the fact that the order of the items in the list isn't important. The following code shows an example of this type of list as you would use it in HTML:

```html
<!-- Shopping list -->
<ul>
    <li>Ice Cream</li>
    <li>Cookies</li>
    <li>Salad</li>
    <li>Soap</li>
</ul>
```

A web page for the preceding code would appear as follows:

- Ice Cream
- Cookies
- Salad
- Soap

Figure 3.7: An unordered list as it appears on a web page

2. Following the unordered list, we have the ordered list, which is expressed in HTML as ol, with li used for the list items. You are probably also quite familiar with this type of list in your everyday life. A common use case for the ordered list could be a recipe shown in a list of sequential steps. With this type of list, the ordering is important, unlike in the unordered list we just looked at. The following code shows an example of this type of list as you would use it in HTML:

```html
<!-- Cheese on toast recipe -->
<ol>
    <li>Place bread under grill until golden brown</li>
```

```
    <li>Flip the bread and place cheese slices</li>
    <li>Cook until cheese is golden brown</li>
    <li>Serve immediately</li>
  </ol>
```

A web page for the preceding code would appear as follows:

1. Place bread under grill until golden brown
2. Flip the bread and place cheese slices
3. Cook until cheese is golden brown
4. Serve immediately

Figure 3.8: An ordered list as it appears on a web page

3. If you want to use an unordered or ordered list but don't want to show bullet points or numbers, respectively, you have a range of options. Using CSS, you can customize the style of the list using the `list-style` property:

```css
/* Alternative styles for unordered lists */
.square {
  list-style-type: square;
}

.circle {
  list-style-type: circle;
}

/* Alternative styles for ordered lists */
.upper-alpha {
  list-style-type: upper-alpha;
}

.upper-roman {
  list-style-type: upper-roman;
}

/* HTML */
<ul class="square">
  <li>Square list style</li>
</ul>

<ul class="circle">
  <li>Circle list style</li>
```

```
</ul>

<ul class="upper-alpha">
  <li>Upper alpha list style</li>
</ul>

<ul class="upper-roman">
  <li>Upper roman list style</li>
</ul>
```

4.   The following figure shows the output of the preceding code:

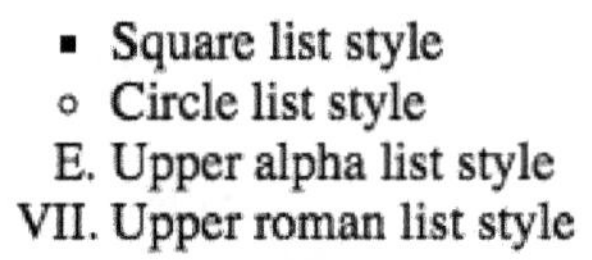

- Square list style
  - Circle list style
- E. Upper alpha list style
- VII. Upper roman list style

Figure 3.9: An unordered list shown with the different list styles

It is also possible to nest lists and use different list styles for each list. In the following HTML code, you can see that we have different lists being nested:

```
<ol>
  <li>Numbered</li>
  <ol class="alphabetic">
    <li>Alphabetic</li>
    <ol class="roman">
      <li>Roman</li>
      <li>Roman</li>
    </ol>
    <li>Alphabetic</li>
  </ol>
  <li>Numbered</li>
</ol>
```

We just need to add two class names to style the two nested lists with alphabetic and Roman list styles, as shown here:

```
.alphabetic {
  list-style: upper-alpha;
}

.roman {
  list-style: upper-roman;
}
```

The following figure shows the output of the preceding code:

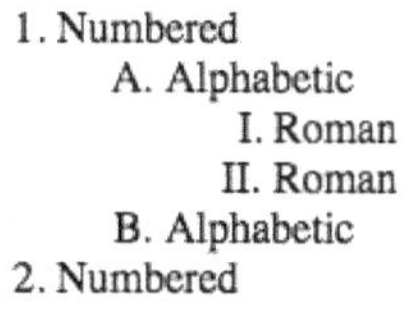

1. Numbered
    A. Alphabetic
        I. Roman
        II. Roman
    B. Alphabetic
2. Numbered

Figure 3.10: Nested lists shown with different list styles

5. The third type of list is the definition list, which is expressed in HTML as dl. Although this type of list is used less frequently than the two other types of lists, you will probably still be familiar with it. The definition list is used when you want to list out pairs of terms and descriptions. The most common use of this type of list is probably dictionary entries. You have the word you are interested in, which is the term, dt, followed by the definition, which is the description, dd. The following is an example of this type of list as you would use it in HTML:

```html
<!-- Dictionary -->
<dl>
  <dt>HTML</dt>
  <dd>Hypertext markup language</dd>

  <dt>CSS</dt>
  <dd>Cascading style sheets</dd>
</dl>
```

A web page for the preceding code would appear as follows:

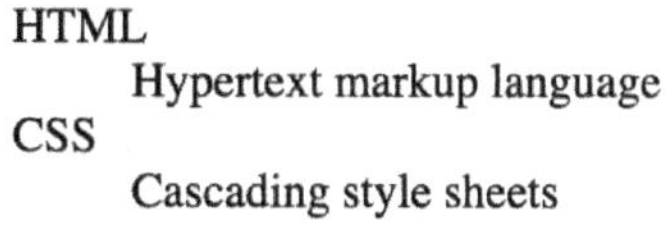

HTML
    Hypertext markup language
CSS
    Cascading style sheets

Figure 3.11: A definition list as it appears on a web page

Exercise 3.1 – combining text-based elements

In this exercise, we will use the following screenshot from the Packt website and write the HTML to match it as closely as possible. This will give us some practice in creating text-based content and recreating the correct HTML for it.

The following figure shows the sample piece of content that we will recreate from the Packt website:

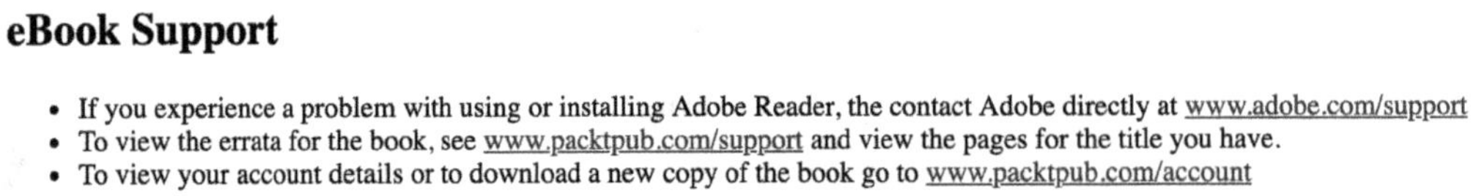

eBook Support

- If you experience a problem with using or installing Adobe Reader, the contact Adobe directly at www.adobe.com/support
- To view the errata for the book, see www.packtpub.com/support and view the pages for the title you have.
- To view your account details or to download a new copy of the book go to www.packtpub.com/account

Figure 3.12: A screenshot from the Packt website

> **Note**
> The complete code for this exercise can be found at `https://packt.link/7W3nk`.

Let's complete the exercise with the following steps:

1. Start by creating a new file in **VS Code** called `text.html`, and use the following code as a starting point:

```
<!DOCTYPE html>
<html>
  <head>
    <title>Combining text based elements</title>
  </head>
  <body>
    <!-- your code will go here -->
  </body>
</html>
```

2. Looking at the preceding screenshot, we can see that we will need a heading. Since the heading is not the top-level heading for the page, we will use h2 in this instance. We will wrap the tag around the text in between the opening and closing body tags, as follows:

```
<body>
  <h2>eBook Support</h2>
</body>
```

3. Below the heading, we have a list of bullet points. We can assume that these are unordered and, hence, use an ul tag for them. Note that each list item contains a link as well, so we need to include an anchor tag for each. We will place our code just below the h2 heading, as follows:

```
<body>
  <h2>eBook Support</h2>
```

```
<ul>
  <li>
    If you experience a problem with using or
    installing Adobe Reader, then contact Adobe
    directly at
    <a href="www.adobe.com/support">
      www.adobe.com/support
    </a>
  </li>
  <li>
    To view the errata for the book, see
    <a href="www.packtpub.com/support">
    www.packtpub.com/support
    </a>
    and view the pages for the title you have.
  </li>
  <li>
    To view your account details or to download a
    new copy of the book, go to
    <a href="www.packtpub.com/account">
      www.packtpub.com/account
    </a>
  </li>
</ul>
</body>
```

If you now right-click on the filename in VS Code on the left-hand side of the screen and select **Open in default browser**, you will see the web page in your browser:

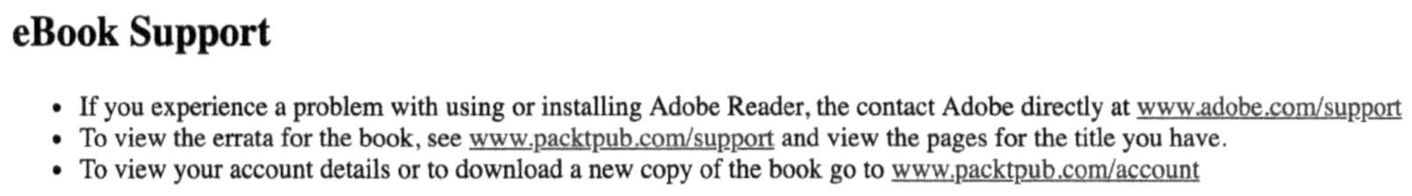

eBook Support

- If you experience a problem with using or installing Adobe Reader, the contact Adobe directly at www.adobe.com/support
- To view the errata for the book, see www.packtpub.com/support and view the pages for the title you have.
- To view your account details or to download a new copy of the book go to www.packtpub.com/account

Figure 3.13: The output of combining text-based elements

You are now getting a feel for the various text-based HTML elements and when you should use them. Before we start looking into how we will go about styling the HTML elements we have just learned about, we will take a look at semantic markup.

Working with new text-based elements

Now, we will take a quick look at some new additions to the HTML language that allow us to further refine our text markup. These elements, while not used as regularly as the previous elements, are useful to know or be aware of.

In this section, we will look at the following HTML text-based elements:

- address
- details
- abbr

address

The address element is used to mark up content that contains contact information. This could be an email address, phone number, physical address, or even a social media profile:

```
<address>
  You can visit us at:<br />
  123 HTML Street
  London
  E1 101
</address>
```

A web page from the preceding code would appear as follows:

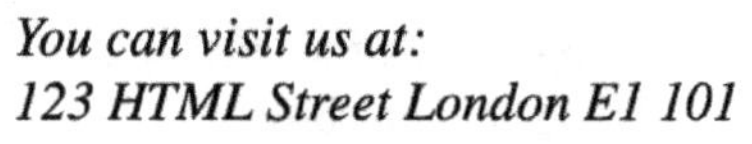

You can visit us at:
123 HTML Street London E1 101

Figure 3.14

Details

The details element creates a widget that allows content to be toggled as visible and hidden to the user. You have to provide a label for the widget using the summary element and provide the hidden content inside of the details element, as shown here:

```
<details>
  <summary>More info</summary>
  More information goes here...
</details>
```

A web page from the preceding code would appear as follows:

▶ More info

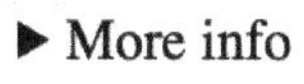

Figure 3.15

And this is what it would look like after clicking the label:

▼ More info
More information goes here…

Figure 3.16

Abbr

The `abbr` element is used to mark up content that represents an abbreviation or acronym. This is used all over the place in the tech world, so it can be a handy element to know about when adding content that includes acronyms or abbreviations, such as the following:

```
<p>
   You are learning to code in
   <abbr>HTML</abbr>
   (HyperText Markup Language) and
   <abbr>CSS</abbr>
   (Cascading Style Sheets)
</p>
```

A web page from the preceding code would appear as follows:

You are learning to code in HTML (HyperText Markup Language) and CSS (Cascading Style Sheets)

Figure 3.17

Now that we have become familiar with some of the newer HTML text-based elements, we will take a look at the concept of "semantic markup," which is essential knowledge you will need when building websites.

Writing semantic markup

You will hear the word "semantic" used often when you read or hear about HTML. The core concept behind semantic markup is to ensure that you use the most meaningful HTML element available to describe the content you mark up. For example, it would be possible for you to wrap the top-level page heading in a `div` tag; however, the `h1` tag conveys the meaning that the content represents – that is, heading level 1. The HTML you write needs to be understandable to both humans and machines and by using the most meaningful element for each piece of content, you improve the meaning of both.

Ensuring that the HTML you write is as semantic as possible also has additional important benefits. The first is that it will make your web pages more easily searchable by search engines. You will also be helping out users who view your websites using a screen reader.

The following code shows some examples of semantic and non-semantic markup:

```
<!-- Semantic markup -->

<h1>I am a top level page heading</h1>
<p>
  This is a paragraph which contains a word with
  <strong>strong</strong> significance
</p>

<!-- Non semantic markup -->
<div>
  I am a top level page heading
</div>
<div>
  This is a paragraph which contains a word with
  <span>strong</span> significance
</div>
```

Hopefully, you now understand the differences between semantic and non-semantic markup. Now that we have some knowledge of the most commonly used HTML elements for content, we can turn to the more fun part of styling.

Styling text-based elements

Until now, we have seen some of the basic text formatting that HTML allows you to implement on a web page. However, depending on the function and the purpose that a web page serves, we might need some styling applied to the text-based elements. Here, we will introduce the common issues surrounding the styling of web page content. We will introduce the different units of measurement, including pixels and relative units. We will then walk through some examples of how to style common web components, such as breadcrumbs and navigation bars.

CSS resets

As you begin styling web pages, you will soon realize that different browsers render your pages slightly differently from each other. This can be very frustrating and makes the task of developing websites that look the same across different browsers a nightmare.

Luckily, there is a well-known solution to alleviate at least some of this frustration. A CSS reset is a style sheet whose sole purpose is to level the playing field across browsers. This file will be loaded before any of your page-specific styles are added. The following is an example of the most basic form of CSS reset code:

```
* {
  margin: 0;
  padding: 0;
}
```

What this will ensure is that all HTML elements will have zero margins and padding before you apply your custom styles to your page. This gets around the issue of different browsers by default adding varying amounts of padding and margin to certain elements.

Although using this reset would be better than having no reset at all, there are more sophisticated CSS resets available. The following shows a popular CSS reset developed by Eric Meyer:

```
/* http://meyerweb.com/eric/tools/css/reset/
   v2.0 | 20110126
   License: none (public domain)
*/

html, body, div, span, applet, object, iframe, h1, h2, h3, h4, h5,
h6, p, blockquote, pre, a, abbr, acronym, address, big, cite, code,
del, dfn, em, img, ins, kbd, q, s, samp, small, strike, strong, sub,
sup, tt, var, b, u, i, center, dl, dt, dd, ol, ul, li, fieldset,
form, label, legend, table, caption, tbody, tfoot, thead, tr, th, td,
article, aside, canvas, details, embed, figure, figcaption, footer,
header, hgroup, menu, nav, output, ruby, section, summary, time, mark,
audio, video {
  margin: 0;
  padding: 0;
  border: 0;
  font-size: 100%;
  font: inherit;
  vertical-align: baseline;
}
/* HTML5 display-role reset for older browsers */
article, aside, details, figcaption, figure, footer,
header, hgroup, menu, nav, section {
```

```
    display: block;
}
body {
  line-height: 1;
}
ol, ul {
  list-style: none;
}
blockquote, q {
  quotes: none;
}
blockquote:before, blockquote:after,
q:before, q:after {
  content: '';
  content: none;
}
table {
  border-collapse: collapse;
  border-spacing: 0;
}
```

As you can see, this is more detailed than our first example and would give us a better chance of reducing cross-browser inconsistencies in our web pages.

CSS text properties

When styling text-based elements with CSS, there are two main groupings of properties you will see used over and over again. These are groups of properties that are based on text and fonts. We will begin by looking at the most commonly used text-based CSS properties that you will need to become familiar with.

The first property we will look at is `color`, which, as the name suggests, is used to set the text color. You will typically set the color value using either hexadecimal, RGB, or a name.

For hexadecimal values, you specify two hexadecimal integers for the colors red, green, and blue. The values you provide range from `00` to `FF`, with `FF` being the most intense version of the specific color and `00` being the least intense. For example, `#FF0000` will be the highest intensity of the color red.

RGB-based colors are different from hexadecimal values because you provide a value between 0 and 255 for each color, with 0 being the lowest intensity and 255 being the maximum intensity.

The following code shows some examples of what this would look like in code form:

```
h1 {
  color: green;
```

```
}

p {
   color: #00ff00;
}

span {
   color: rgb(0, 255, 0);
}
```

We will learn more about text colors and background colors in *Chapter 7*.

You may have noticed that all the text we have seen so far aligns to the left by default. With CSS, we have the power to change this using the `text-align` property. You can style your text to be left-aligned, centered, or right-aligned. The following shows some examples of what this would look like in code form:

```
p {
   text-align: center;
}
```

A web page for the preceding code would appear as follows:

> Last night of all, When yond same star that's westward from the pole Had
> made his course t' illume that part of heaven Where now it burns,
> Marcellus and myself, The bell then beating one

Figure 3.18: Centrally aligned text as it appears on a web page

---

Note

The text used to explain the CSS text properties is a work of Shakespeare from `https://www.opensourceshakespeare.org/views/plays/play_view.php?WorkID=hamlet&Act=1&Scene=1&Scope=scene`.

If you want to underline some text, you can use the `text-decoration` property, which gives you the ability to strike through text as well:

```
.underline {
   text-decoration: underline;
}
```

A web page for the preceding code would appear as follows:

Last night of all, When yond same star that's westward from the pole Had made his course t' illume that part of heaven Where now it burns, Marcellus and myself, The bell then beating one

Figure 3.19: Underlined text as it appears on a web page

```
.line-through {
  text-decoration: line-through;
}
```

A web page for the preceding code would appear as follows:

~~Last night of all, When yond same star that's westward from the pole Had made his course t' illume that part of heaven Where now it burns, Marcellus and myself, The bell then beating one~~

Figure 3.20: Strikethrough text as it appears on a web page

Another common styling requirement for text is the ability to control how it is capitalized. For this, we have the `text-transform` property, which gives you the ability to transform the text. By default, the text is set to lowercase, but with this property, you can set the text to all caps or title case. The following code shows how you would use this property:

```
.uppercase {
  text-transform: uppercase;
}
```

A web page for the preceding code would appear as follows:

LAST NIGHT OF ALL, WHEN YOND SAME STAR THAT'S WESTWARD FROM THE POLE HAD MADE HIS COURSE T' ILLUME THAT PART OF HEAVEN WHERE NOW IT BURNS, MARCELLUS AND MYSELF, THE BELL THEN BEATING ONE

Figure 3.21: Uppercase text as it appears on a web page

```
.lowercase {
  text-transform: lowercase;
}
```

A web page for the preceding code would appear as follows:

last night of all, when yond same star that's westward from the pole had
made his course t' illume that part of heaven where now it burns, marcellus
and myself, the bell then beating one

Figure 3.22: Lowercase text as it appears on a web page

```
.capitalize {
  text-transform: capitalize;
}
```

A web page for the preceding code would appear as follows:

Last Night Of All, When Yond Same Star That's Westward From The Pole
Had Made His Course T' Illume That Part Of Heaven Where Now It Burns,
Marcellus And Myself, The Bell Then Beating One

Figure 3.23: Title-case text as it appears on a web page

Finally, on our tour of the most used text-based CSS properties, we have the line-height property,
which is used to control the amount of vertical spacing between lines of text. This is a property you
will see used over and over again, as different types of copy will require different line heights. We will
see two extreme examples of how small and large line-height values affect the readability of the text:

```
.small-line-height {
  line-height: .5;
}
```

A web page for the preceding code would appear as follows:

Lorem ipsum dolor sit amet, consectetur adipiscing elit.
Suspendisse odio felis, euismod eu est ac, facilisis euismod
odio. Mauris sed vestibulum neque.

Figure 3.24: A small line-height as it appears on a web page

```
.large-line-height {
  line-height: 1.5;
}
```

A web page for the preceding code would appear as follows:

Lorem ipsum dolor sit amet, consectetur adipiscing elit.
Suspendisse odio felis, euismod eu est ac, facilisis euismod
odio. Mauris sed vestibulum neque.

Figure 3.25: A large line-height as it appears on a web page

CSS font properties

The second group of CSS properties we are concerned with is font-based properties. These are responsible for defining the font family, the size of the font, and the weight. We will now take a quick tour of the most commonly used CSS font properties that you should become familiar with.

The first property we will look at is the font-family property, which, as you might have guessed, sets the font family. When using this property, you will usually provide a list of different font families in order of priority. If the web browser doesn't support your first choice of font, it will default to the second or find a font family it can load. The following code shows how this property can be used:

```
body {
    font-family: "Times New Roman", Times, serif;
}
```

The browser will attempt to load the "Times New Roman", font from the user's computer first. If it cannot load the font, then it will try "Times" and if this fails, it will load the generic "serif" font family.

Now that we know how to set the correct font family for our text, we will need to control the size. For this, we can use the font-size property. You can set the value of the size using pixels or relative units, such as ems. The following shows examples of this property's use:

```
/* pixels */
h1 {
    font-size: 50px;
}

p {
    font-size: 16px;
}

/* ems */
h1 {
    font-size: 3.125em;
```

```
}

p {
    font-size: 16px;
}
```

The benefit of using ems for the unit of measurement is that it allows the user to control the font size. By default, a browser's font size is set to 16 px, but if a user wants to increase their default font size, they can. With em units, the font sizes will scale according to the base font size. With pixels used as units, you give the user less flexibility in controlling font sizes.

The last font-based CSS property we will look at is the `font-weight` property. This is used to control the weight of a font, typically to make a font bold, as by default, the weight is set to normal. The following shows you how this would look in code:

```
span {
    font-weight: bold;
}
```

We now have all the knowledge we require to build a realistic-looking web page. We will put this theory into practice by building components for web pages one by one.

The display Property

Before we move on to the next exercise, we need to look into a new CSS property called `display`. By default, elements are either set to `block` or `inline`. Now, `block` elements will take up all horizontal space, while `inline` elements only take up as much horizontal space as their content. An example of a block-level element is a `div` tag, and an example of an `inline` element is `span`. Sometimes, you need to style a `block` element as an `inline` element, and vice versa. You can do this by using the following CSS:

```
div {
    display: inline;
}

span {
    display: block;
}
```

The video store product page (revisited)

Remember the video store product page examples from *Chapter 2*, where we stepped through several CSS layout techniques? We are going to be using a more detailed version of this page to work through some exercises, demonstrating how to code some of the key components on a page. By doing this, we should be able to put most of the theory from this chapter into practice. The following figure is the revised wireframe for the page:

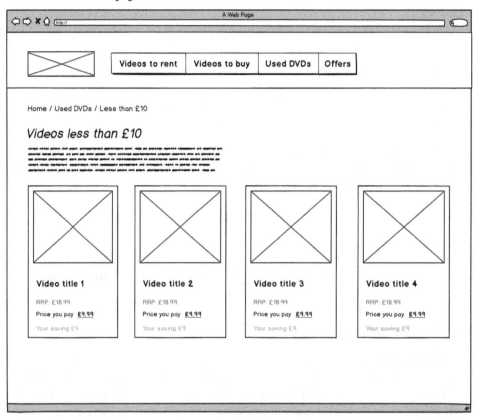

Figure 3.26: The video store product page revisited

Exercise 3.2 – navigation

In this exercise, we will step through the process of writing the HTML and CSS for the navigation component shown in the preceding wireframe. The following figure shows the navigation component in more detail:

Figure 3.27: Navigation

> **Note**
>
> The complete code for this exercise can be found at `https://packt.link/RnVPJ`.

The following are the steps to complete this exercise:

1. First, let's create a file called `nav.html` in VS Code using the following HTML code as a starter file:

```
<!DOCTYPE html>
<html>
  <head>
    <title>Exercise 3.02</title>
    <style>
      /* your CSS will go here */
    </style>
  </head>
  <body>
    <!-- your HTML will go here -->
  </body>
</html>
```

2. Next, we need to decide on how we should semantically mark up this component. We will start with the nav tag, which we learned about in the previous chapter. We will place this in between the opening and closing body tags, as follows:

```
<body>
  <nav></nav>
</body>
```

3. For the actual list of navigation links, using `ul` is most appropriate, as the order isn't of any significance. We will place the code for `ul` in between our `nav` tags, as follows:

```
<body>
  <nav>
    <ul>
      <li>Videos to rent</li>
      <li>Videos to buy</li>
      <li>Used DVDs</li>
      <li>Offers</li>
    </ul>
  </nav>
</body>
```

4. Since users of the web page will want to be able to click on each of the navigation items, we must add anchors for each item. We will place the anchors around the text contained in our list items, as follows:

```
<body>
  <nav>
    <ul>
      <li><a href="">Videos to rent</a></li>
      <li><a href="">Videos to buy</a></li>
      <li><a href="">Used DVDs</a></li>
      <li><a href="">Offers</a></li>
    </ul>
  </nav>
</body>
```

5. We will now have a look in the browser at what this will look like by default without any custom styling. To do this, right-click on the filename in VS Code on the left-hand side of the screen and select **open in default browser**. You should see something that resembles the following screenshot:

- Videos to rent
- Videos to buy
- Used DVDs
- Offers

Figure 3.28: Default navigation

6. This is very different from what we want the navigation to look like. To correct this, we will now add some CSS to more closely match the navigation component shown in the wireframe. We will start by adding a basic CSS reset, which will be placed in between the opening and closing `style` tags, as follows:

```
<style>
  {
    margin: 0;
    padding: 0;
  }
</style>
```

7. We will use the `flex` layout technique we learned about in *Chapter 2*. We will also remove the default bullets shown to the left of the list items:

```
<style>
  {
```

```
    margin: 0;
    padding: 0;
  }

  nav ul {
    display: flex;
    list-style: none;
  }
</style>
```

8. We will now add some styling to the links so that they are no longer blue and more closely resemble the links shown in the wireframe. We will make the text bold, remove the default underline that gets applied to the links, and give the link some padding:

```
nav a {
  color: black;
  font-weight: bold;
  display: block;
  padding: 15px;
  text-decoration: none;
}
```

The following figure shows the output of the web page so far:

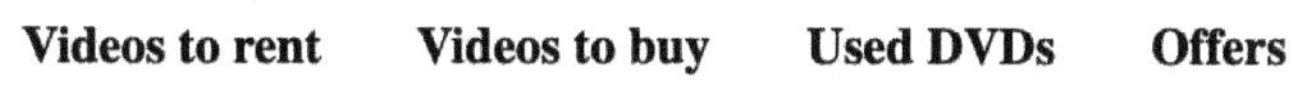

Videos to rent   Videos to buy   Used DVDs   Offers

Figure 3.29: The output without hover

9. Finally, to apply the hover selector on the anchor elements and set the link text to underline when a user hovers over the text, we will add the following code:

```
nav a:hover {
  text-decoration: underline;
}
```

10. If you now right-click on the filename in VS Code on the left-hand side of the screen and select **open in default browser**, you will see the navigation component in your browser.

You should now see something that looks like the following screenshot:

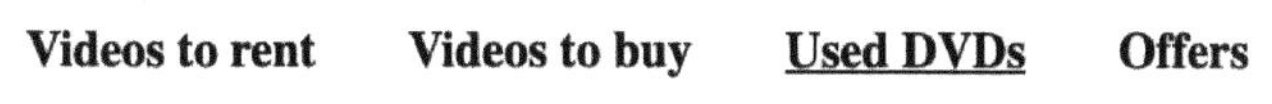

Videos to rent   Videos to buy   <u>Used DVDs</u>   Offers

Figure 3.30: Styled navigation

Breadcrumbs

On websites that have lots of pages, it is common for pages to contain what is called breadcrumbs. This is a list of links that easily allow a user to see the context of the current page within the website structure and easily navigate to the parent pages. The following screenshot shows a breadcrumb taken from the Packt website at `https://www.packtpub.com/`

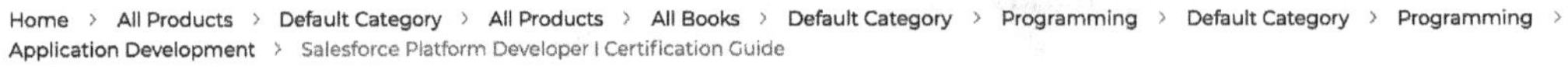

Home > All Products > Default Category > All Products > All Books > Default Category > Programming > Default Category > Programming > Application Development > Salesforce Platform Developer I Certification Guide

Figure 3.31: A breadcrumb from the Packt web page

As you can see in the preceding figure, web pages can contain many parent pages or categories. The breadcrumbs of the current page are for a book in the Packt online store. Moving from right to left, the breadcrumb links get more specific, moving from the home page to the closest parent page. The user is free to click on any link, which will take them back to a page related to the current page.

Exercise 3.3 – breadcrumbs

In this exercise, we will step through the process of writing the HTML and CSS for the breadcrumb component shown in *Figure 3.22*. The following figure shows it in more detail:

Home / Used DVDs / Less than £10

Figure 3.32: A breadcrumb

> **Note**
> The complete code for this exercise can be found at `https://packt.link/fWsaJ`.

The following steps show how to complete a breadcrumb component similar to the preceding example:

1. First, let's create a file called `breadcrumb.html` in VS Code using the following HTML code as a starter file:

```
<!DOCTYPE html>
<html>
  <head>
    <title>Exercise 3.03</title>
    <style>
      /* your CSS will go here */
    </style>
  </head>
```

```
<body>
  <!-- your HTML will go here -->
</body>
</html>
```

2. Next, we need to decide what the best HTML tag will be for the breadcrumb. Since this is a list of links and the ordering is important, we will use ol. We will place this in between the body tags, as follows:

```
<body>
  <ol class="breadcrumb">
    <li>Home</li>
    <li>Used DVDs</li>
    <li>Less than £10</li>
  </ol>
</body>
```

3. We will then add anchors to all but the last item, as that represents the current page the user is viewing so doesn't need to be clickable:

```
<body>
  <ol class="breadcrumb">
    <li><a href="">Home</a></li>
    <li><a href="">Used DVDs</a></li>
    <li>Less than £10</li>
  </ol>
</body>
```

4. Let's now take a look at what this will look like in the browser:

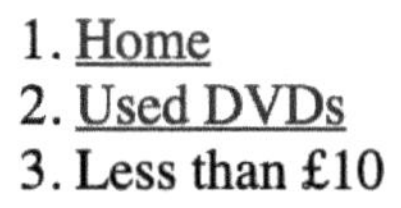

1. <u>Home</u>
2. <u>Used DVDs</u>
3. Less than £10

Figure 3.33: The default breadcrumb

5. We will now start adding our styling. Again, we will start with a basic CSS reset, and we will add the styles in between the opening and closing style tags, as follows:

```
<style>
  {
    margin: 0;
    padding: 0;
  }
</style>
```

6. We will use the `flex` layout technique again, which we learned about in the previous chapter, and remove the default numbers from the ordered list:

```
<style>
  {
    margin: 0;
    padding: 0;
  }

  .breadcrumb {
    display: flex;
    list-style: none;
  }
</style>
```

7. Let's style the list items in the ordered list. We will first add some padding to the list of items. Then, we will add a forward slash at the end of all the list items except the last one. We will add some margin to the left to ensure that our list items are nicely separated:

```
.breadcrumb li {
  padding: 10px;
}

.breadcrumb li:after {
  content: '/';
  margin-left: 20px;
}

.breadcrumb li:last-child:after {
  content: '';
}
```

8. Finally, we will style the anchors, making sure the color of the text is black and only showing an underline when a user hovers over the link:

```
.breadcrumb a {
  color: black;
  text-decoration: none;
}

.breadcrumb a:hover {
  text-decoration: underline;
}
```

9. If you now right-click on the filename in VS Code on the left-hand side of the screen and select **open in default browser**, you will see the breadcrumb component in your browser.

You should now see something similar to the following figure in your browser:

Home   /   Used DVDs   /   Less than £10

Figure 3.34: A styled breadcrumb

Exercise 3.4 – the page heading and introduction

We will now write the HTML and CSS for the heading and introduction section of the wireframe. The following figure shows it in more detail:

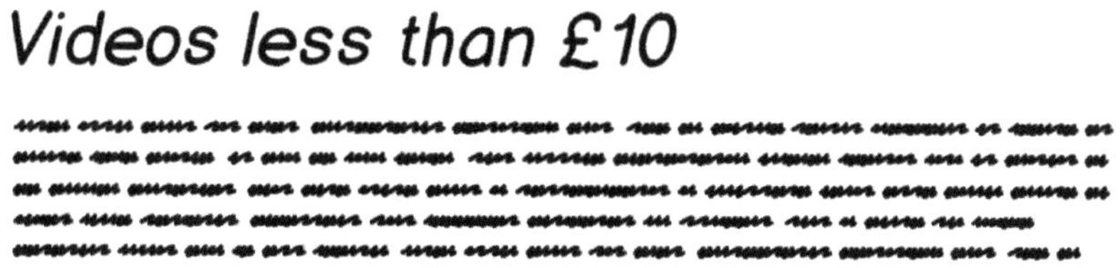

Videos less than £10

Figure 3.35: The introduction section

> **Note**
> The complete code for this exercise can be found at `https://packt.link/JcamP`.

The steps to complete the exercise are as follows:

1. First, let's create a file called `text.html` in VS Code using the following HTML code as a starter file:

```
<!DOCTYPE html>
<html>
  <head>
    <title>Exercise 3.04</title>
    <style>
      /* your CSS will go here */
    </style>
  </head>
  <body>
```

```
        <!-- your HTML will go here -->
      </body>
    </html>
```

2. Then, let's decide on the correct markup to use. Since the heading represents the top-level heading for the page, we will use h1 and then a plain p for the introduction. We will place the HTML in between the opening and closing body tags:

```
<body>
  <section class="intro">
  <h1>Videos less than £10</h1>
  <p>
     Lorem ipsum dolor sit amet, consectetur adipiscing
     elit. In bibendum non purus quis vestibulum.
     Pellentesque ultricies quam lacus, ut tristique
     sapien tristique et.
  </p>
  </section>
</body>
```

3. Let's now take a look at what this looks like in the browser:

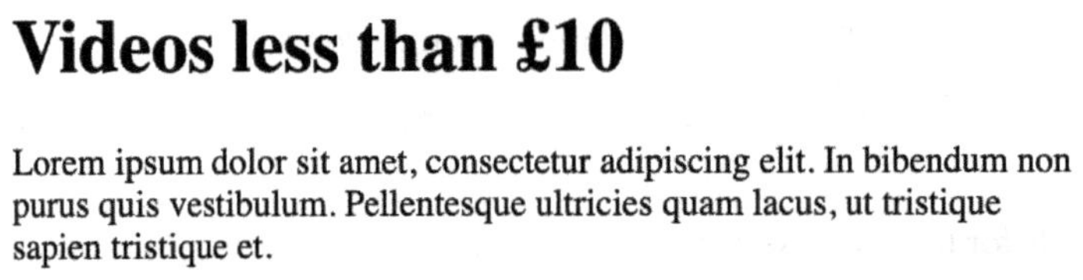

Videos less than £10

Lorem ipsum dolor sit amet, consectetur adipiscing elit. In bibendum non purus quis vestibulum. Pellentesque ultricies quam lacus, ut tristique sapien tristique et.

Figure 3.36: The default introduction section

4. This is actually pretty close to how we want these elements to look. We will just adjust the heading margin and the line height for the paragraph to make the text more readable. We will place the CSS in between the opening and closing style tags, as follows:

```
<style>
  .intro {
    margin: 30px 0;
    padding-left: 10px;
    width: 50%;
  }

  .intro h1 {
    margin-bottom: 15px;
  }
```

```
  .intro p {
    line-height: 1.5;
  }
</style>
```

5. If you now right-click on the filename in VS Code on the left-hand side of the screen and select **open in default browser**, you will see the heading and paragraph.

You should now see something similar to the following figure in your browser:

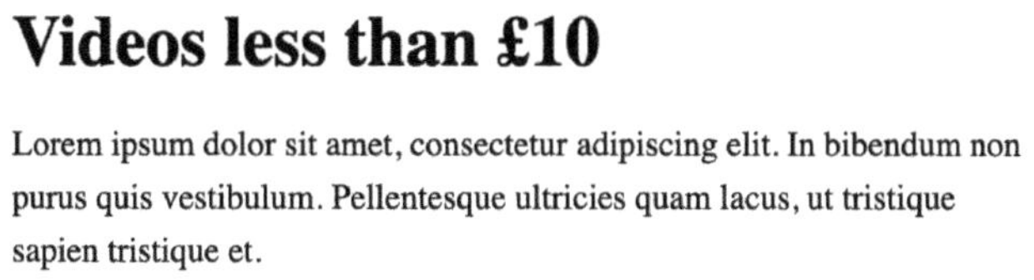

Videos less than £10
Lorem ipsum dolor sit amet, consectetur adipiscing elit. In bibendum non purus quis vestibulum. Pellentesque ultricies quam lacus, ut tristique sapien tristique et.

Figure 3.37: A styled introduction section

Exercise 3.5 – product cards

In this exercise, we will step through the process of writing the HTML and CSS for the product card component shown in the wireframe. The following figure shows the product card in more detail:

Video title 1
RRP: £18.99
Price you pay £9.99
Your saving £9

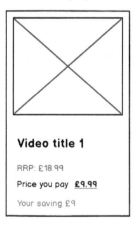

Figure 3.38: The product card

> **Note**
>
> The complete code for this exercise can be found at `https://packt.link/KatBg`.

The steps to complete the exercise are as follows:

1. First, let's create a file called `product.html` in VS Code using the following starter file:

```html
<!DOCTYPE html>
<html>
  <head>
    <title>Exercise 3.05</title>
    <style>
      /* your CSS will go here */
    </style>
  </head>
  <body>
    <!-- your HTML will go here -->
  </body>
</html>
```

2. Next, let's use a `div` tag with a class name for the outer wrapper of the component. We will place this in between the opening and closing `body` tags, as follows:

```html
<body>
  <div class="product-card">

  </div>
</body>
```

3. Now, let's add an image tag; we will add the image URL and some `alt` text:

```html
<body>
  <div class="product-card">
    <img
      src="https://dummyimage.com/300x300/7EC0EE/
        000&text=Product+Image+1"
      alt="Product image 1" />
  </div>
</body>
```

4. We will then use h2 for the heading. Note how we wrap the text in an anchor so that the user will be able to click on the card:

```html
<body>
  <div class="product-card">
    <img
      src="https://dummyimage.com/300x300/7EC0EE/
        000&text=Product+Image+1"
```

```
      alt="Product image 1" />
    <h2><a href="">Video title 1</a></h2>
  </div>
</body>
```

5. Now, we will add the markup for the pricing information. Note that we will add a class name to allow us to style the information individually:

```
<body>
  <div class="product-card">
    <img
      src="https://dummyimage.com/300x300/7EC0EE/
        000&text=Product+Image+1"
      alt="Product image 1" />
    <h2><a href="">Video title 1</a></h2>
    <p class="original-price">RRP: £18.99</p>
    <p class="current-price">
      Price you pay <span>£9.99
      </span>
    </p>
    <p class="saving">Your saving £9</p>
  </div>
</body>
```

6. Let's now take a look at what this looks like in the browser without any styling added:

Video title 1

RRP: £18.99

Price you pay £9.99

Your saving £9

Figure 3.39: The default product card

7. Now, let's add some styling for the product card container in between the `style` tags, as follows. Note how we give the product card a black `border` and some `padding`:

```
<style>
  .product-card {
    display: inline-block;
    border: 1px solid black;
    padding: 15px;
  }
</style>
```

8. Then, we will add some styling for the individual elements of the product card. Starting with the image element, we will ensure the width of the image stretches to 100% by adding the following code:

```
.product-card img {
  width: 100%;
}
```

We will add a margin to the level 2 header using the following code:

```
.product-card h2 {
  margin: 30px 0 15px;
}
```

The following code styles the links:

```
.product-card a {
  color: black;
  text-decoration: none;
}
```

With the help of the following code, we will style the paragraph element:

```
.product-card p {
  line-height: 1.5;
}
```

We will add the following code to style the original price, current price, and the savings, as per the wireframe shown in *Figure 3.34*:

```
.original-price {
  color: grey;
  text-transform: uppercase;
}
.current-price span {
  font-weight: bold;
```

```
    text-decoration: underline;
}
.saving {
    color: green;
}
```

If you now right-click on the filename in VS Code on the left-hand side of the screen and select **open in default browser**, you will see the product component in your browser.

You should now see something similar to the following figure in your browser:

Figure 3.40: A styled product card

Exercise 3.6 – putting it all together

Now that we have built the individual parts of the product page, we have the fun task of putting them all together to assemble a web page. We will be able to reuse the code we have written already and just need to make minor tweaks to the CSS to get the page looking good. Our aim is to produce a web page that resembles the wireframe shown in the following figure:

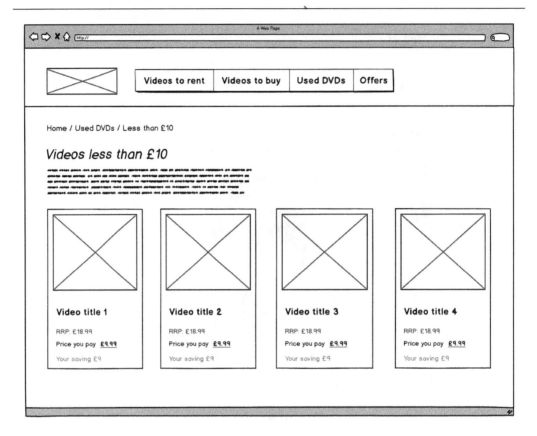

Figure 3.41: The wireframe of the expected output

> **Note**
>
> The complete code for this exercise can be found at https://packt.link/bHkEr.

The steps to create the preceding product page are as follows:

1. First, let's create a file called product-page.html in VS Code. We will use the following HTML page template; note that we will use an inline style sheet to make things easier:

```
<!DOCTYPE html>
<html>
<head>
  <title>Video product page</title>
  <style type="text/css">
    {
```

```
      margin: 0;
      padding: 0;
    }
  body {
    font-family: sans-serif;
    margin: 0 auto;
    width: 1200px;
  }
  header {
    align-items: center;
    display: flex;
    margin-bottom: 25px;
  }
  nav {
    margin-left: 30px;
  }
  .product-cards {
    display: grid;
    grid-template-columns: auto auto auto;
    margin-bottom: 30px;
  }
  /* your styles will go here */
  </style>
</head>
<body>
  <header>
    <img
      src="https://dummyimage.com/200x100/000/
        fff&text=Logo"
      alt=""
    />
    <!-- navigation will go here -->
  </header>
  <section>
    <!-- breadcrumb will go here -->
  </section>
  <!-- introduction section will go here -->
  <section class="product-cards">
    <!-- product cards will go here -->
  </section>
</body>
</html>
```

2. Now, let's add the navigation CSS, which is the same as the code used in *Exercise 3.2 – navigation*, to the following HTML from the same exercise:

```
<nav>
  <ul>
    <li><a href="">Videos to rent</a></li>
    <li><a href="">Videos to buy</a></li>
    <li><a href="">Used DVDs</a></li>
    <li><a href="">Offers</a></li>
  </ul>
</nav>
```

3. Then, we will add the styles for the breadcrumb component, which are the same ones used in *Exercise 3.3 – breadcrumbs*, to the following HTML from the same exercise:

```
<section>
  <ol class="breadcrumb">
    <li><a href="">Home</a></li>
    <li><a href="">Used DVDs</a></li>
    <li>Less than £10</li>
  </ol>
</section>
```

4. We will then add the introduction section's CSS, which is the same as that of *Exercise 3.4 – the page heading and introduction*, to the following HTML from the same exercise:

```
<section class="intro">
  <h1>Videos less than £10</h1>
  <p>
    Lorem ipsum dolor sit amet, consectetur
    adipiscing elit. In bibendum non purus quis
    vestibulum. Pellentesque ultricies quam lacus,
    ut tristique sapien tristique et.
  </p>
</section>
```

5. Finally, we will add the product card's CSS, which is the same as that of *Exercise 3.5 – product cards*, to the following HTML from the same exercise:

```
<section class="product-cards">
  <div class="product-card">
    <img
      src="https://dummyimage.com/300x300/000/
        fff&text=Product+Image+1"
      alt=""
    />
```

```
            <h2><a href="">Video title 1</a></h2>
            <p class="original-price">RRP: £18.99</p>
            <p class="current-price">
              Price you pay <span>£9.99</span>
            </p>
            <p class="saving">Your saving £9</p>
          </div>
    <!-- Similar to the above product card 1 that is video title 1,
    the product cards for video titles 2, 3, and 4 can be copied
    from Exercise 3.05, Product Cards -->
        </section>
```

6.   If you now right-click on the filename in VS Code on the left-hand side of the screen and select **open in default browser**, you will see the web page.

You should now see something similar to the following figure in your browser:

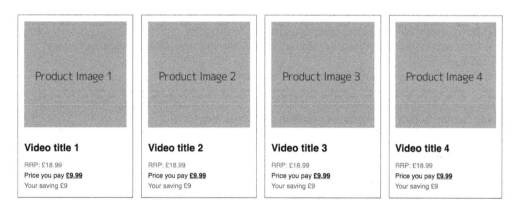

Figure 3.42: A styled product page

Activity 3.1 – converting a newspaper article to a web page

Get a copy of a recent newspaper, choose a particular article, and note down what HTML elements would be used if the paper were a web page. Then, create a web page version of the newspaper article, using semantic markup and CSS to recreate the layout as closely as possible:

1. Get a copy of a newspaper article and annotate it with a pen to label the individual HTML elements.

2. Create a file named `home.html` in VS Code. You can use the starter HTML from a previous exercise as a starting point.

3. Start writing out the HTML for the news article.

4. Finally, style the text and layout using CSS.

An example of how you could annotate a newspaper article to distinguish the different page elements can be seen in the following figure:

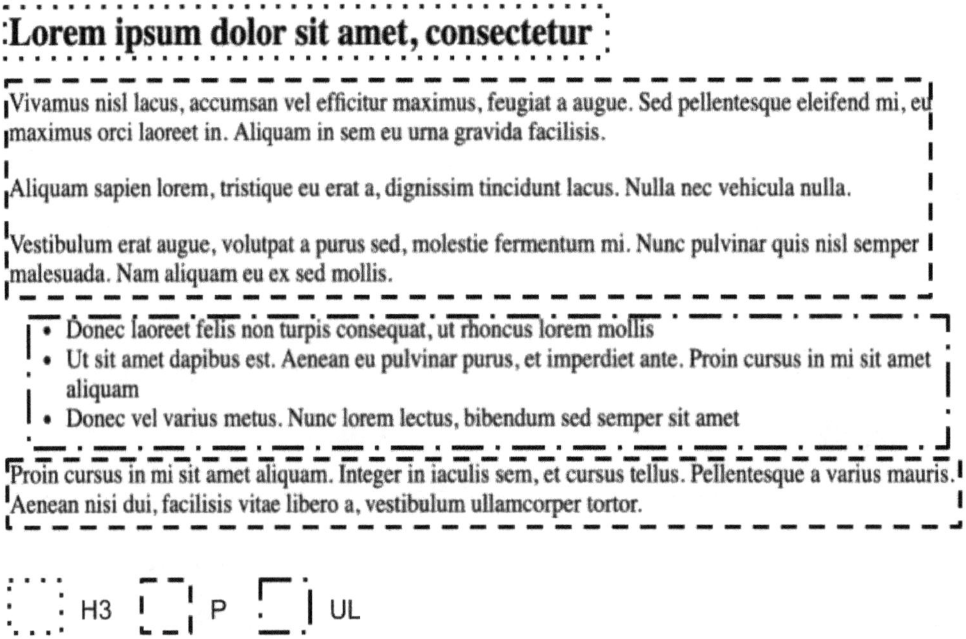

Lorem ipsum dolor sit amet, consectetur

Vivamus nisl lacus, accumsan vel efficitur maximus, feugiat a augue. Sed pellentesque eleifend mi, eu maximus orci laoreet in. Aliquam in sem eu urna gravida facilisis.

Aliquam sapien lorem, tristique eu erat a, dignissim tincidunt lacus. Nulla nec vehicula nulla.

Vestibulum erat augue, volutpat a purus sed, molestie fermentum mi. Nunc pulvinar quis nisl semper malesuada. Nam aliquam eu ex sed mollis.

- Donec laoreet felis non turpis consequat, ut rhoncus lorem mollis
- Ut sit amet dapibus est. Aenean eu pulvinar purus, et imperdiet ante. Proin cursus in mi sit amet aliquam
- Donec vel varius metus. Nunc lorem lectus, bibendum sed semper sit amet

Proin cursus in mi sit amet aliquam. Integer in iaculis sem, et cursus tellus. Pellentesque a varius mauris. Aenean nisi dui, facilisis vitae libero a, vestibulum ullamcorper tortor.

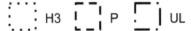

H3   P   UL

Figure 3.43: An annotated article sample

> **Note**
> The solution to this activity is available on GitHub at `https://packt.link/6D8aL`.

Summary

In this chapter, we continued our journey into building web pages. We first looked at the most common text-based HTML elements, such as headings, paragraphs, and lists. We then looked into the most common styling methods available for text-based content. To put this new knowledge into practice, we then walked through building a complete web page.

We took some time to understand the concept and importance of writing semantic HTML. We were also introduced to some common web page components, such as navigation and breadcrumbs.

In the next chapter, we will learn how to take our web pages to the next level. We will learn how to make our web pages far more interesting by adding videos, animation, and forms.

Part 2:
Understanding Website Fundamentals

In this part, we'll explore the fundamental principles of creating interactive web applications. We'll start by learning how to create and style forms, laying a strong foundation for user interaction. Next, we'll explore techniques for animating web pages to enhance the user experience with smooth, intuitive flows. Finally, we'll delve into themes, color schemes, and polishing techniques to elevate the visual appeal of your website, all while optimizing performance.

This section contains the following chapters:

- *Chapter 4, Creating and Styling Forms*
- *Chapter 5, Adding Animation to Web Pages*
- *Chapter 6, Themes, Color, and Polishing Techniques*

4

Creating and Styling Forms

In the previous chapters, we studied how to build web pages that contain static text-based content. From this chapter onward, we will learn how to make web pages much more interesting, starting with forms.

Forms allow users to actually interact with a website. They enable users to sign up for services, order products online, and so on. Forms are arguably one of the most crucial aspects of business websites, as without forms, no transactions can take place online. Businesses require online forms to capture user details when new user accounts are created, allowing users to select flight details when booking a holiday online, for instance. Without forms, many online businesses would not be able to function. With this in mind, developing complex forms is an essential skill to add to your toolbelt as a web developer.

In this chapter, we will take a look at the most common elements that are used to build forms with HTML. These HTML elements include text inputs, radio buttons, checkboxes, text areas, and submit buttons. Once we've gained an understanding of the most commonly used form elements, we will look at styling concerns. This will include techniques to make our form elements look visually appealing to a wide range of users. We will put all of this into practice by building different online forms.

The following topics will be covered in the chapter:

- Learning common form-based HTML elements
- Learning common styling techniques for these form-based elements
- Learning how to make forms accessible

By the end of this chapter, you will be able to use the correct HTML form elements to build an online form, customize form elements to improve the look and feel of your web forms, build online forms, apply form validation styles, and identify when to use checkboxes over radio buttons.

This chapter introduces HTML forms and associated elements used within forms. We will first look at the most common HTML form elements used when building forms. We will then take a look at some common techniques for styling forms. We will then put all of this into practice by building signup and checkout forms for a video store.

Technical requirements

The code files for this chapter can be found at `https://packt.link/Aq9iN`.

Introducing form elements

HTML provides us with a variety of elements that are used to build forms. While browsing the web, you may have noticed that online forms typically have similar elements. Most forms will contain input fields such as text inputs, checkboxes, and select boxes.

In this section, we will look at the following HTML form elements:

- `form`
- `input`
- `label`
- `textarea`
- `fieldset`
- `select`
- `button`

form

The first element we need to know about when creating forms is the `form` element. This is the outermost element, which contains all other form elements, such as inputs and buttons. The `form` element requires you to pass two attributes, which are the `action` and `method` attributes. The `action` attribute allows a developer to specify the URL that the form data will go to after it has been submitted. The `method` attribute allows the developer to specify whether the form data should be sent via `get` or `post`. You will typically use the `get` method when you deal with unsecured data, since it will be present in a query string. Conversely, the `post` method is typically used when you deal with secure data or a large amount of soft data. The following code snippet shows an example of what an empty form would look like in HTML:

```html
<form action="url_to_send_form_data" method="post">
  <!-- form elements go here -->
</form>
```

input

Now that we have our `form` element, we can start building a form using different elements. The first one we will look at is the `input` element. This is the element you would use when creating text input fields, radio buttons, and checkboxes. The `input` element is the most important of all the form elements we will look at, and you will find yourself using it over and over again. The `input` element requires two attributes, which are `type` and `name`. The `type` attribute is used to specify what type of input you want, such as radio buttons, checkboxes, or text. The `name` attribute gives the element a unique name that is required when submitting the form. This is so that the form's data can be organized into **key-value** pairs with a unique `name`, corresponding to a `value`. It should be noted that the order in which you add the attributes has no significance. The following code snippet shows how to create a text field using the `input` element:

```
<!-- text input -->
<form action="url_to_send_form_data" method="post">
  <div>
    First name: <br />
    <input type="text" name="firstname" />
  </div>
  <div>
    Last name: <br />
    <input type="text" name="lastname" />
  </div>
</form>
```

The following figure shows the output for the preceding code:

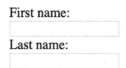

Figure 4.1: Text inputs, as shown in the browser

Sometimes, when creating text inputs, you will want to limit the number of characters a user can add. A common example of this is when you want to restrict the number of characters for a new username in account signup forms. You can use the `maxlength` attribute and set the maximum number of characters allowed for the `input` field. The following code snippet shows how you would use this attribute:

```
<input type="text" name="username" maxlength="20" />
```

There is also a specialist type of text input that is solely for email addresses. To create an email input, you simply set the `type` to `"email"`. This input type has built-in validation that checks whether the input text is a valid email address. The following code snippet shows how to create an email input:

```
<!-- email input -->
<form action="url_to_send_form_data" method="post">
```

```
  <div>
    Email: <br />
    <input type="email" name="email"/>
  </div>
</form>
```

The following figure shows the output for the preceding code:

Email:

john@smith.com

Figure 4.2: The email input, as shown in the browser

There is a type of text input that is used solely for passwords. To create a password input, you simply set the type to "password". This input type will mask the text entered by the user to hide the **password** text. The following code snippet shows how to create a password input:

```
<!-- password input -->
<form action="url_to_send_form_data" method="post">
  <div>
    Password: <br />
    <input type="password" name="password"/>
  </div>
</form>
```

The following figure shows the output for the preceding code:

Password:

••••••••

Figure 4.3: Password input, as shown in the browser

When using checkboxes, you will give all of them a unique value for the name attribute, and you will need to give each checkbox a unique value attribute, as shown in the following code:

```
<!-- checkboxes -->
<form action="url_to_send_form_data" method="post">
  <div>
    <input type="checkbox" name="color1" value="red" /> Red
  </div>
  <div>
    <input type="checkbox" name="color2" value="green" />
    Green
  </div>
```

```
<div>
  <input type="checkbox" name="color3" value="blue" />
    Blue
</div>
</form>
```

The following figure shows the output for the preceding code:

Figure 4.4: Checkboxes, as shown in the browser

With checkboxes, you can select multiple values at a time. A common use case for checkboxes is when selecting multiple filters to search results.

When using **radio** buttons, you will give all of them the same value for the name attribute, since only one value can be selected. However, you will need to give each radio button a unique value attribute, as shown in the following code snippet:

```
<!-- radio buttons -->
<form action="url_to_send_form_data" method="post">
  <div>
    <input type="radio" name="color" value="red" /> Red
  </div>
  <div>
    <input type="radio" name="color" value="green" /> Green
  </div>
  <div>
    <input type="radio" name="color" value="blue" /> Blue
  </div>
</form>
```

The following figure shows the output for the preceding code:

Figure 4.5: Radio buttons, as shown in the browser

In contrast to checkboxes, with radio buttons, the user can select only one value. A common use case for radio buttons is selecting a delivery option when ordering online.

To learn more about the input control, head over to the MDN documentation to see a more comprehensive overview: `https://developer.mozilla.org/en-US/docs/Web/HTML/Element/input`.

label

Now that we know how to create text inputs, checkboxes, and radio buttons, we need to look at the `label` element. In the previous examples, you might have noticed that we had text associated with the input fields either before or after an `input` element. The `label` element allows us to associate a piece of text with a `form` element and select the `form` element by clicking on the text. If we were to just include some text, as we did in *Figure 4.1*, we would lose this benefit and make our form less accessible for screen reader users, since there would not be an associated `label` to call out when presenting a form element. The `label` element has an attribute called `for`, which we need to give the `id` of the element we wish to associate the label with. The following code snippet shows this in action:

```
<!-- text inputs with labels -->
<form action="url_to_send_form_data" method="post">
  <div>
    <label for="first_name">First name:</label><br />
    <input type="text" name="firstname" id="first_name" />
  </div>
  <div>
    <label for="last_name">Last name:</label><br />
    <input type="text" name="lastname" id="last_name" />
  </div>
</form>
```

The following figure shows the output for the preceding code:

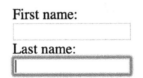

Figure 4.6: Text inputs with the labels, as shown in the browser

textarea

Imagine you are creating a "comments" section for a community web page. You might want a user to comment on a video or a blog post. However, using text input is not ideal for long text messages. In such scenarios, when you want to allow the user to add more than one line of text, you can use the `textarea` element to capture larger amounts of text. You can specify the size of `textarea` with the `rows` and `cols` attributes. The following code snippet shows how to include `textarea` within a form:

```
<!-- textarea -->
<form action="url_to_send_form_data" method="post">
  <div>
    <label for="first_name">First name:</label><br />
    <input type="text" name="firstname" id="first_name" />
  </div>
  <div>
    <label for="last_name">Last name:</label><br />
    <input type="text" name="lastname" id="last_name" />
  </div>
  <div>
    <label for="message">Message:</label><br />
    <textarea id="message" rows="5" cols="20"></textarea>
  </div>
</form>
```

The following figure shows the output for the preceding code:

Figure 4.7: textarea, as shown in the browser

fieldset

HTML provides us with a semantic tag to group related form elements, which is called the `fieldset` element. This element is mostly used with larger forms when you want to group related form elements together. You will probably have used online forms that make use of the `fieldset` element without realizing it. A common use case is when you have a large form with a section for personal details and a section for delivery details. Both of these sections of the form would be wrapped in a `fieldset` element. The following shows how we could include more than one form using `fieldset`:

```
<!-- fieldset -->
<form action="url_to_send_form_data" method="post">
  <fieldset>
    <div>
```

```
      <label for="first_name">First name:</label><br />
      <input
        type="text"
        name="firstname"
        id="first_name"
      />
  </div>
  <div>
    <label for="last_name">Last name:</label><br />
    <input type="text" name="lastname" id="last_name" />
  </div>
  <div>
    <label for="message">Message:</label><br />
    <textarea id="message" rows="5" cols="20"></textarea>
  </div>
  </fieldset>
  <p>Do you like HTML?</p>
  <fieldset>
    <div>
      <input type="radio" id="yes" name="yes">
      <label for="yes">Yes</label>
    </div>
    <div>
      <input type="radio" id="no" name="no">
      <label for="no">No</label>
    </div>
  </fieldset>
</form>
```

The following figure shows the output for the preceding code:

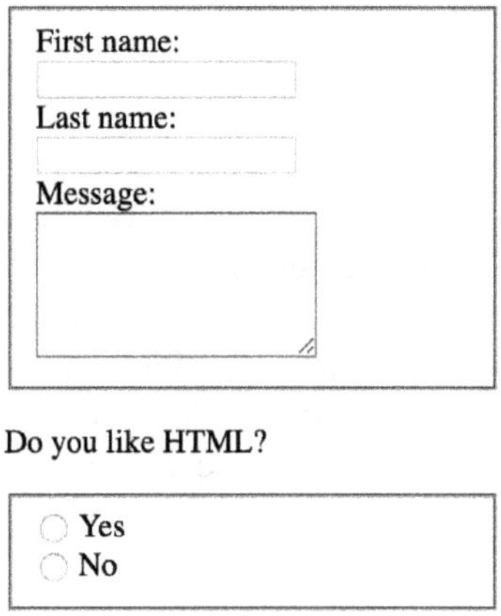

Figure 4.8: A form with fieldset, as shown in the browser

select

HTML provides us with the `select` element to create select boxes. These are typically used when you have a long list of options and you want a user to select only one. Some common examples include lists of countries, addresses, and years of birth. Inside the `select` element, you provide a list of options inside of an `option` element. The following shows an example of how this looks in HTML:

```
<!-- select -->
<form action="url_to_send_form_data" method="post">
  <fieldset>
    <label for="countries">Country:</label><br />
    <select id="countries">
      <option value="england">England</option>
      <option value="scotland">Scotland</option>
      <option value="ireland">Ireland</option>
      <option value="wales">Wales</option>
    </select>
  </fieldset>
</form>
```

The following figure shows the output for the preceding code:

Figure 4.9: The select box closed, as shown in the browser

By clicking on the blue arrows on the right-hand side of the select box, we can display the options in the following figure:

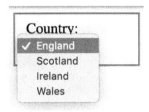

Figure 4.10: The select box open, as shown in the browser

button

Now that we finally have a range of form elements we can use to build web forms, we just need to know how to submit a form. The button element requires a type attribute that can have three different values – firstly, the button value, which has no default behavior; the "reset" value, which, once clicked, will reset all form values; and finally, the "submit" value, which will submit the form once clicked. For this, we will use the button element and give a value of "submit" in the type attribute:

```
<button type="submit">Submit</button>
Or
<input type="button" value="Submit" />
```

Exercise 4.1 – creating a simple form

In this exercise, we will write HTML to create a simple web form. Our aim is to produce the following web form:

Figure 4.11: A simple web form, as shown in the browser

> **Note**
> The complete code for this exercise can be found at https://packt.link/PGpEh.

Let's complete the exercise by following the following steps:

1. Start by creating a new file in VS Code called `simple-form.html`, using the following code as your starting point:

```
<!DOCTYPE html>
<html>
<head>
  <title>Simple form</title>
</head>
<body>
  <h1>Create new account</h1>
  <form action="url_to_send_form_data" method="post">
    <fieldset>
      <!-- your code will go here -->
    </fieldset>
  </form>
</body>
</html>
```

2. Then, we will add the HTML for the `title` and `first` name fields between the opening and closing `fieldset` elements:

```
<fieldset>
  <div>
    <label for="title">Title:</label><br />
    <select id="title">
      <option value="Mr">Mr</option>
      <option value="Mrs">Mrs</option>
      <option value="Ms">Ms</option>
      <option value="Miss">Miss</option>
    </select>
  </div>
  <div>
    <label for="first_name">First name:</label><br />
    <input
      type="text"
      name="firstname"
      id="first_name"
    />
  </div>
</fieldset>
```

3. Next, we will add the HTML for the `last` name field. We will add this after the closing `div` element, which wraps the HTML for the `first` name field:

```
<div>
  <label for="first_name">First name:</label><br />
  <input
    type="text"
    name="firstname"
    id="first_name"
  />
</div>
<div>
  <label for="first_name">First name:</label><br />
  <input
    type="text"
    name="firstname"
    id="first_name"
  />
</div>
<div>
  <label for="last_name">Last name:</label><br />
  <input
    type="text"
    name="lastname"
    id="last_name"
  />
</div>
```

4. Now, we will add HTML for the `message` field. Note how this will be a `textarea` element instead of an `input` element to allow the user to enter more than one line of text:

```
<div>
  <label for="first_name">First name:</label><br />
  <input
    type="text"
    name="firstname"
    id="first_name" />
</div>
<div>
  <label for="last_name">Last name:</label><br />
  <input
    type="text"
    name="lastname"
    id="last_name"
```

```
      />
    </div>
    <div>
      <label for="message">Message:</label><br />
      <textarea
        id="message"
        rows="5"
        cols="20">
      </textarea>
    </div>
```

5. Finally, we need to add the submit button, as follows:

```
    <div>
      <label for="first_name">First name:</label><br />
      <input
        type="text"
        name="firstname"
        id="first_name"
      />
    </div>
    <div>
      <label for="last_name">Last name:</label><br />
      <input
        type="text"
        name="lastname"
        id="last_name"
      />
    </div>
    <div>
      <label for="message">Message:</label><br />
      <textarea
        id="last_name"
        rows="5"
        cols="20">
      </textarea>
    </div>
    <button type="submit">Submit</button>
```

If you now right-click on the filename in VS Code, on the left-hand side of the screen, and select **Open in default browser**, you will see the form in your browser.

You should now have a form that looks like the following figure:

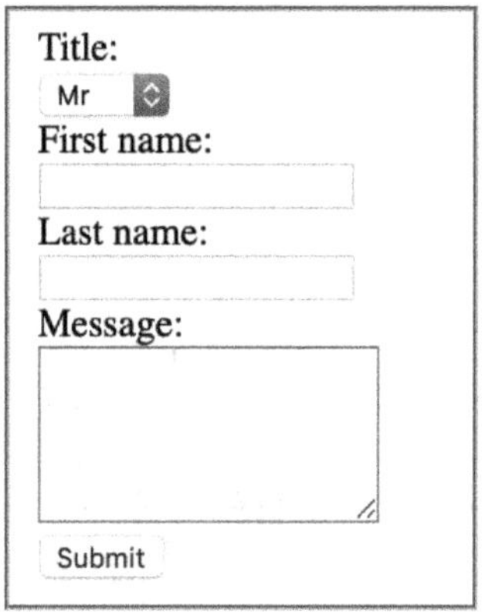

Figure 4.12: The Submit button, as shown in the browser

Now that we have become acquainted with the most commonly used HTML form elements, we will now look at how to style them.

Styling form elements

In the examples in the previous section, the forms do not look very visually appealing by default, but luckily, we can improve the look and feel of our forms using CSS.

In this section, we will look at styling the following:

- Textboxes
- Textareas
- Labels
- Buttons
- Select boxes
- Validation styling

Labels, textboxes, and textareas

The first form elements we will look at styling are the label, textboxes, and `textarea` elements. These are probably the most common form elements, and it is very straightforward to improve the look and feel of these elements with minimal code.

To style labels, you will typically just adjust the font you use and the size of the text. It is common to have the label element sit either on top of its associated form element or to the left.

For textboxes and textareas, typically, you will be interested in changing the size of these elements. It is common to remove the default border around these elements. Another common stylistic addition to textboxes and textareas is to add a placeholder attribute that provides the user with some text, helping them decide what needs to be typed into the textbox or textarea.

To illustrate an example of how to style these elements, we will start with the following markup, noting the addition of placeholder attributes:

```html
<!-- HTML -->
<form action="url_to_send_form_data" method="post">
  <div>
    <label for="first_name">First name:</label><br />
    <input
      type="text"
      name="firstname"
      id="first_name"
      placeholder="Your first name"
    />
  </div>
  <div>
    <label for="last_name">Last name:</label><br />
    <input
      type="text"
      name="lastname"
      id="last_name"
      placeholder="Your last name"
    />
  </div>
  <div>
    <label for="message">Message:</label><br />
    <textarea
      id="last_name"
      rows="5"
      cols="20"
      placeholder="Your message">
    </textarea>
  </div>
</form>


/* CSS */
:root {
  --border-color: #666;
}
```

```
*  {
  font-family: arial,sans-serif;
}

label {
  font-size: 20px;
}

div {
  margin-bottom: 30px;
}

input,
textarea {
  border: 0;
  border-bottom: 1px solid var(--border-color);
  padding: 10px 0;
  width: 200px;
}
```

In the preceding CSS, note that we have applied a font family to all text elements. We have set the label text size to 20 px and added a bottom margin to the `div` elements so that the form elements are nicely spaced, vertically. Finally, we have removed the default border applied to the `input` and `textarea` elements, replacing it with just a **border** on the bottom.

With just minimal CSS, we have improved the look and feel of our form drastically, as can be seen in the following screenshot:

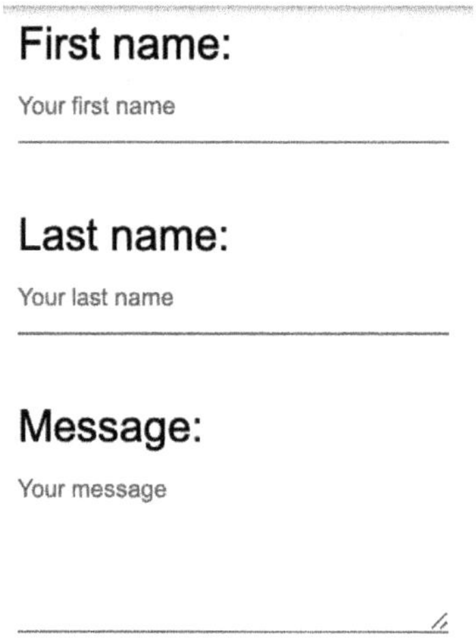

First name:

Your first name

_____

Last name:

Your last name

_____

Message:

Your message

_____

Figure 4.13: Styled labels, textboxes, and textarea elements

Buttons

We will now look into styling the buttons that are used to submit a web form. Typically, you will see buttons with various background colors and different sizes applied when viewing websites with forms. Out of the box, the `button` element looks pretty ugly, so you will rarely see buttons without some CSS applied to them. The following is an example of how you could style a `submit` button:

```
<!-- HTML -->
<button type="submit">Submit</button>


/* CSS */
:root {
  --bg-color: #999;
  --bg-active-color: #888;
  --text-color: #fff;
}

button {
  background: var(--bg-color);
  border: 0;
  color: var(--text-color);
  cursor: pointer;
  font-size: 12px;
  height: 50px;
  width: 200px;
  text-transform: uppercase;
}

button:hover {
  background: var(--bg-active-color);
}

button:active {
  background: var(--bg-color);
}
```

The preceding CSS sets a background color, removes the border that is added to buttons by default, applies some styling to the button text, and finally, sets a width and height:

Figure 4.14: A styled submit button

Select boxes

The last form element we will look at for styling is the `select` box. Typically, this is styled to make the `select` box look similar to a **textbox** within a form. It is common for developers to add a custom-styled downward-pointing arrow to the right-hand side of the `select` box. The following is an example of how you could style a `select` box:

```html
<!-- HTML -->
<div class="select-wrapper">
  <select id="countries">
    <option value="england">England</option>
    <option value="scotland">Scotland</option>
    <option value="ireland">Ireland</option>
    <option value="wales">Wales</option>
  </select>
</div>


/* CSS */
:root {
  --border-color: #666;
}

select {
  background: transparent;
  border: 0;
  border-radius: 0;
  border-bottom: 1px solid var(--border-color);
  box-shadow: none;
  color: var(--border-color);
  padding: 10px 0;
  width: 200px;
  -webkit-appearance: none;
}

.select-wrapper {
  position: relative;
  width: 200px;
}

.select-wrapper:after {
  content: '< >';
```

```
    color: var(--border-color);
    font-size: 14px;
    top: 8px;
    right: 0;
    transform: rotate(90deg);
    position: absolute;
    z-index: -1;
}
```

The preceding CSS contains styling that essentially overrides what a `select` box looks like in the browser by default. Firstly, we need to remove the default background color, cancel the border, and apply just the bottom border. We also remove the custom `box-shadow` property, which is also applied to select boxes by default. Finally, to add a custom `select` box icon, we use the `after` **pseudo** selector to add the `'< >'` characters:

Figure 4.15: A styled select box

Validation styling

In real-world scenarios, simply formatting and styling a form appropriately is not enough. As a web user, you may encounter cases where form validation is performed before submitting a form. For example, while registering on a website, a user may accidentally submit a form before it is filled in completely or submit an incorrectly filled-in form. Validation styling comes into play when you want to highlight the fact that a form is incomplete or incorrectly filled in.

You will probably have experienced form validation on web forms you have used in the past. HTML provides us with a `required` attribute, which we can apply to any form elements that we require input for. The `required` attribute plays an important role in contact forms – for example, on the Packt website's contact form (`https://www.packtpub.com/en-us/help/contact`), note that the name and email fields are required, and the user cannot submit the form until a value for each is added.

This is in contrast with some form elements where the input is optional. With CSS, we can use the `:valid` and `:invalid` pseudo **selectors** to style elements based on valid or invalid form values. We will now do an exercise that will walk us through an example of validation styles in action.

Exercise 4.02 – creating a form with validation styling

In this exercise, we will develop a simple web form that contains some **validation** styling. Our aim is to produce a web form like the one shown in the following figure:

First name:
John
_____

Last name:
Smith
_____

Country:
England   ◊
_____

Message:
Hello there...

_____
                                                           ⁄⁄

┌───┐
│   SUBMIT   │
└───┘

Figure 4.16: The expected output

┌──┐
│ **Note**   │
│ The complete code for this exercise can be found at `https://packt.link/q1PLm`. │
└──┘

Let's complete the exercise with the following steps:

1. Start by creating a new file in VS Code called `validation-form.html`, and use the following code as your starting point:

```
<!DOCTYPE html>
<html>
<head>
  <title>Validation form</title>
  <style>
```

```
      body {
        font-family: arial, sans-serif;
      }
    </style>
  </head>
  <body>
    <form action="url_to_send_form_data" method="post">
      <fieldset>
        <!-- your code will go here -->
      </fieldset>
    </form>
  </body>
</html>
```

2. We will now add the HTML for the first name and last name form fields between the opening and closing fieldset tags. Note how we have added required attributes to both of the input elements:

```
<fieldset>
  <div>
    <label for="first_name">First name:</label><br />
    <input
      type="text"
      name="firstname"
      id="first_name"
      placeholder="Your first name" required
    />
  </div>
  <div>
    <label for="last_name">Last name:</label><br />
    <input
      type="text"
      name="lastname"
      id="last_name"
      placeholder="Your last name" required
    />
  </div>
  <div>
    <label for="last_name">Country:</label><br />
    <div class="select-wrapper">
  <select id="countries">
    <option value="england">England</option>
    <option value="scotland">Scotland</option>
```

```
        <option value="ireland">Ireland</option>
        <option value="wales">Wales</option>
    </select>
  </div>
</div>
<div>
    <label for="message">Message:</label><br />
    <textarea
      id="last_name"
      rows="5"
      cols="20"
      placeholder="Your message">
    </textarea>
  </div>
<div>
<button type="submit">Submit</button>
</div>
</fieldset>
```

3. Now, we will turn to the CSS. We will first add some styling, which will deal with **spacing** the `div` and `fieldset` elements:

```
div {
  margin-bottom: 30px;
}

fieldset {
  border: 0;
  padding: 30px;
}
```

4. Next, we will style the individual form elements one by one. The label's font size is set to 20 px, and the styling for the `input` and `textarea` elements is the same, as shown in the following code snippet:

```
label {
  font-size: 20px;
}

input,
textarea {
  border: 0;
  border-bottom: 1px solid grey;
  padding: 10px 0;
  width: 200px;
}
```

With respect to the expected output as shown in *Figure 4.16*, we style `select`, as shown in the following code snippet:

```css
select {
  background: transparent;
  border: 0;
  border-radius: 0;
  border-bottom: 1px solid grey;
  box-shadow: none;
  color: #666;
  -webkit-appearance: none;
  padding: 10px 0;
  width: 200px;
}
```

We will use the following snippet of code to complete styling `select`:

```css
.select-wrapper {
  position: relative;
  width: 200px;
}

.select-wrapper:after {
  content: '<>';
  color: #666;
  font-size: 14px;
  top: 8px;
  right: 0;
  transform: rotate(90deg);
  position: absolute;
  z-index: -1;
}
```

To style the button, we will use the styling, as shown in the following code snippet:

```css
button {
  background: #999;
  border: 0;
  color: white;
  font-size: 12px;
  height: 50px;
  width: 200px;
  text-transform: uppercase;
}
```

5.   Finally, we will add styles to validate the form elements that have a `required` attribute:

```
:root {
  --valid-color: green;
  --invalid-color: red;
}

input:valid,
textarea:valid {
  border-bottom-color: var(--valid-color);
}

input:invalid,
textarea:invalid {
  border-bottom-color: var(--invalid-color);
}
```

If you now right-click on the filename in VS Code, on the left-hand side of the screen, and select **open in default browser**, you will see the form in your browser. When you try to submit an incomplete form that has validation in it, it will not submit, and you will see something like the following screenshot:

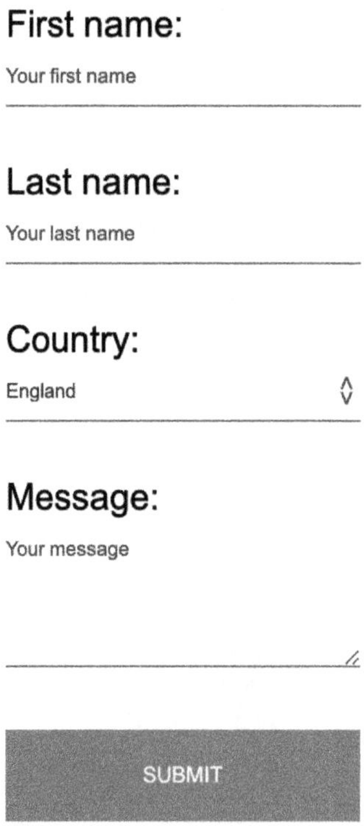

First name:

Your first name

_____

Last name:

Your last name

_____

Country:

England   ⌃⌄

_____

Message:

Your message

_____

SUBMIT

Figure 4.17: A screenshot of the resultant form

We have just looked into ways of styling HTML forms, including validation styling. Next, we will take a look at how we can incorporate HTML tables for the layout of our forms.

Using tables for form layout

Now, we will take a quick look at using HTML tables for the layout of forms. We will look at the following HTML table elements:

- table
- tr
- td

table

The `table` element is used to create a table structure in HTML composed of columns and rows. To create a table, you first need to do this:

```
<table>
   … rows and columns will go here
</table>
```

tr

Then, we need the `tr` element to create a row for the table:

```
<table>
   <tr>
   … column will go here
   </tr>
</table>
```

td

Then, we need the `td` element to create a column for the table:

```
<table>
   <tr>
     <td>content here</td>
   </tr>
</table>
```

A table-based form layout

Now, let's combine using a table for layout with a basic form, as we saw in a previous example in this chapter:

```
<p>Do you like HTML?</p>
<table>
  <form>
    <tr>
      <td>
        <input type="radio" id="yes">
      </td>
      <td>
        <label for="yes">Yes</label>
      </td>
    </tr>
    <tr>
      <td>
        <input type="radio" id="no">
      </td>
      <td>
        <label for="no">No</label>
      </td>
    </tr>
  </form>
</table>
```

The following figure shows the output for the preceding code:

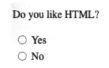

Figure 4.18: A table-based form, as shown in the browser

We have just looked at ways of using HTML tables to lay out our forms. Next, we will take a look at how we can make our forms accessible so that they can be used by the widest possible range of users.

Making forms accessible

Accessibility is a growing concern for websites. It is vital we make our websites as accessible as possible for all users. When working with forms in HTML, there are a few key areas in which we can improve accessibility, which we will now look at:

- Labeling form elements

- Grouping related form elements
- Validating user input

Labeling form elements correctly

We have already used labels for form elements, which is one of the most crucial ways of ensuring form accessibility. Whenever you use a `label` element, you need to ensure you use the `for` attribute, as follows:

```
<label for="first_name">First name:</label>
<input id="first_name" type="text" name="first_name" />
```

This ensures that you explicitly associate a descriptive label with a form control.

Grouping related form elements

We previously grouped our form elements with the `fieldset` element, but there is an additional element called `legend` that is useful for this purpose. Grouping form elements improves accessibility by breaking down a form into smaller focused parts. It is easier for a user to understand each part of the form at a time. Using a `legend` element allows you to provide a description of what the purpose of the form is, grouped by a `fieldset` element. An example of this is shown here:

```
<fieldset>
  <legend>Favorite web language?</legend>
  <div>
    <input type="radio" id="html" name="html" />
    <label for="html">HTML</label>
  </div>
  <div>
    <input type="radio" id="css" name="css" />
    <label for="css">CSS</label>
  </div>
</fieldset>
```

The following figure shows the output for the preceding code:

Figure 4.19: A legend and fieldset form, as shown in the browser

Validating user input

Validating form fields is another great and easy way to improve our form's accessibility. We improve form accessibility by helping our users avoid mistakes when inputting data into our form fields. Luckily, this is easy to achieve, using the `required` attribute for fields that we require user input from. This is for situations where we want to ensure that the data entered is in a certain format – for example, email addresses, website URLs, or dates. We can make use of the `type` attribute when using text inputs:

```
<label for="first_name">First name(required)</label>
<input
  type="text"
  id="first_name"
  name="first_name"
  required
/>
```

The following figure shows the output for the preceding code if a user tried to submit a form without entering any text into the text field:

Figure 4.20: The required attribute, as shown in the browser

The following shows some examples of making use of the `type` attribute when using text inputs in which you want user input to be for a certain format (e.g., an email address):

```
<label for="email">Email</label>
<input type="email" id="email" name="email" />
```

For a full list of the available `type` attributes. visit `https://developer.mozilla.org/en-US/docs/Learn/Forms/HTML5_input_types`.

To learn more about web accessibility, visit `https://developer.mozilla.org/en-US/docs/Web/Accessibility`.

Video store forms

To put our newfound knowledge into practice, we will now build two complex forms for the video store project page examples from the previous chapter, where we built a whole web page, component by component.

Exercise 4.03 – a new account signup form

In this exercise, we will write some HTML and CSS to create an account signup form. Our aim is to produce a web form like the following wireframe:

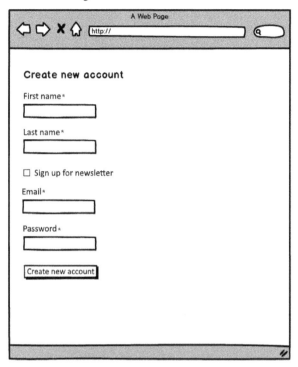

Figure 4.21: A wireframe of a new account form

Note

The complete code for this exercise can be found at https://packt.link/sGtB9.

Let's complete the exercise by following the following steps:

1. Start by creating a new file in VS Code called signup-form.html, and use the following code as your starting point:

```
<!DOCTYPE html>
<html>
<head>
  <title>Signup form</title>
```

```
  <style>
    body {
      font-family: arial, sans-serif;
    }
  </style>
</head>
<body>
  <h1>Create new account</h1>
  <form action="url_to_send_form_data" method="post">
    <fieldset>
    </fieldset>
  </form>
</body>
</html>
```

2. Now, we will add the HTML for the form up until just before the **Sign up for newsletter** checkbox. Note how we have added the `required` attribute to the elements that we require the user to provide a value for:

```
<fieldset>
  <div>
    <label for="first_name">
      First name: <span>*</span>
    </label>
    <input
      id="first_name"
      type="text"
      name="firstname"
      required
    />
  </div>
  <div>
    <label for="last_name">
      Last name: <span>*</span>
    </label>
    <input
      id="last_name"
      type="text"
      name="lastname"
      required
    />
  </div>
</fieldset>
```

3. Next, we will add the checkbox to sign up to the newsletter. In order to have the checkbox first and the description later, `label` comes after `input`:

```
<fieldset>
  <div>
    <label for="first_name">
      First name: <span>*</span>
    </label>
    <input
      id="first_name"
      type="text"
      name="firstname"
      required
    />
  </div>
  <div>
    <label for="last_name">
      Last name: <span>*</span>
    </label>
    <input
      id="last_name"
      type="text"
      name="lastname"
      required
    />
  </div>
  <div class="checkbox">
    <input
      id="newsletter"
      type="checkbox"
      name="newsletter"
      value="yes"
    />
    <label for="newsletter">
      Sign up for newsletter
    </label>
  </div>
</fieldset>
```

4. Now, we will add the remaining fields and the `submit` button below the checkbox shown in the **wireframe**. Again, make sure to add the `required` attribute to elements that need a value to be provided:

```
<div class="checkbox">
  <input
    id="newsletter"
```

```
    type="checkbox"
    name="newsletter"
    value="yes"
  />
  <label for="newsletter">
    Sign up for newsletter
  </label>
</div>
<div>
  <label for="email">Email: <span>*</span></label>
  <input
    id="email"
    type="email"
    name="email"
    required
  />
</div>
<div>
  <label for="password">
    Password: <span>*</span>
  </label>
  <input
    id="password"
    type="password"
    name="password"
    required
  />
</div>
<button type="submit">Create new account</button>
```

5. If you look at the HTML file in your web browser, you should now have a web page that looks like the following screenshot:

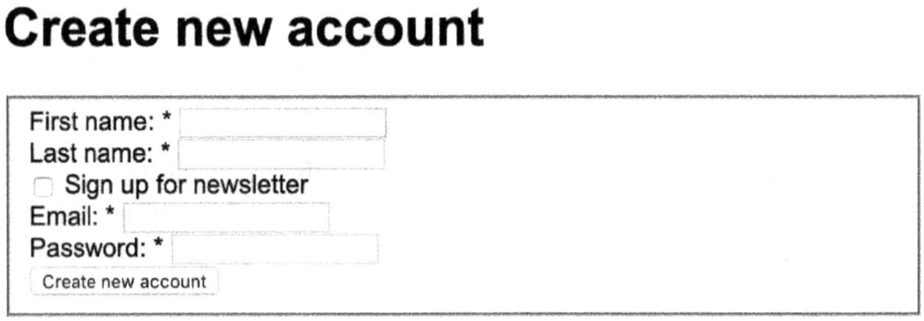

Create new account

First name: *
Last name: *
☐ Sign up for newsletter
Email: *
Password: *
Create new account

Figure 4.22: An unstyled signup form in the browser

6. We can now add the CSS to style the form so that it more closely resembles the one shown in the preceding wireframe. We will start by applying some styles that deal with the spacing of elements and then style the red asterisks shown in the wireframe. The following code snippet will be inside the `style` tag, following the font style applied to the body element:

```css
fieldset {
  border: 0;
}
fieldset>div {
  margin-bottom: 30px;
}
label {
  display: block;
  margin-bottom: 10px;
}
label span {
  color: red;
}
input {
  padding: 10px;
  width: 200px;
}
```

7. Next, we will apply some styles to the `valid` and `invalid` states for the text inputs. Feel free to change the colors of these styles to suit your own taste:

```css
input:valid {
  border: 2px solid green;
}

input:invalid {
  border: 2px solid red;
}
```

8. Next, we will add some styling for the checkbox and the `submit` button, as shown in the following code. Again, feel free to adjust the colors to suit your needs:

```css
.checkbox input ,
.radio input {
  float: left;
  margin-right: 10px;
  width: auto;
}
button {
  background: green;
```

```
    border: 0;
    color: white;
    width: 224px;
    padding: 10px;
    text-transform: uppercase;
}
```

If you now right-click on the filename in VS Code, on the left-hand side of the screen, and select **open in default browser**, you will see the form in your browser.

You should now have a form that looks like the following figure:

Create new account

First name: *

John

Last name: *

Smith

☑ Sign up for newsletter

Email: *

john@smith.com

Password: *

••••••••

CREATE NEW ACCOUNT

Figure 4.23: A styled signup form in the browser

9. We will now check that the email validation works correctly for the form by adding some text to the email field that isn't a valid **email** address:

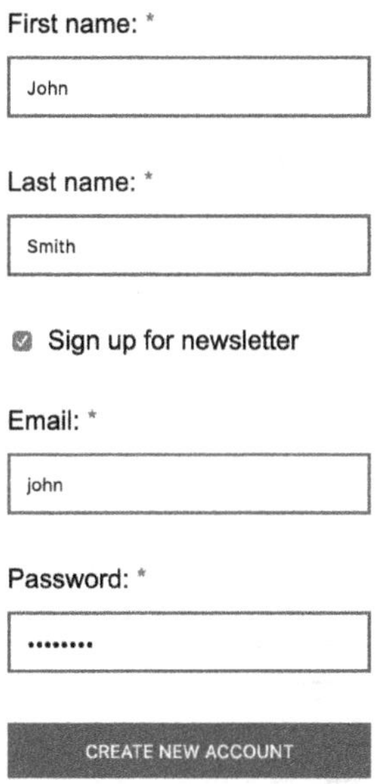
First name: *
John
Last name: *
Smith
Sign up for newsletter
Email: *
john
Password: *
••••••••
CREATE NEW ACCOUNT

Figure 4.24: A form showing an invalid email address

10. Next, we will check all of the required field validations by trying to submit the form without adding input to any of the required fields:

First name: *
Last name: *
Sign up for newsletter
Email: *
Password: *
CREATE NEW ACCOUNT

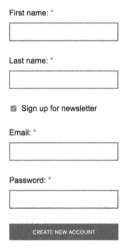

Figure 4.25: A form highlighting fields that need to be filled in to submit the form

We have now completed the signup form for our video store. We will also need to create a checkout page to allow a user to complete their online purchases. The next exercise will show how you can create a checkout form.

Exercise 4.04 – a checkout form

In this exercise, we will write some HTML and CSS to create a checkout form. We will make use of all the concepts and form elements we have learned so far in the chapter. Our aim is to produce a form similar to the following figure:

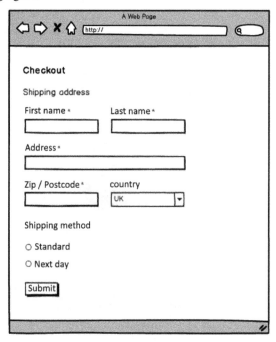

Figure 4.26: A checkout form wireframe

> **Note**
>
> The complete code for this exercise can be found at https://packt.link/xvl1y.

Let's complete the exercise by following the following steps:

1. Start by creating a new file in VS Code called checkout-form.html, and use the following code as a starting point. This gives us the skeletal HTML and CSS we will need to get started writing the checkout form:

```
<!DOCTYPE html>
<html>
<head>
```

```
<title>Checkout form</title>
<style>
  body {
    font-family: arial, sans-serif;
  }
</style>
</head>
<body>
  <h1>Checkout</h1>
  <form action="url_to_send_form_data" method="post">
    <fieldset>
    </fieldset>
  </form>
</body>
</html>
```

2. Now, we will add the HTML for the form inside `fieldset` up until the `address` field. Note how we have added the `required` attribute to the elements that we require the user to provide a value for:

```
<h2>Shipping address</h2>
<div class="double-input">
  <div>
    <label for="first_name">
      First name: <span>*</span>
    </label>
    <input
      id="first_name"
      type="text"
      name="firstname"
      required
    />
  </div>
  <div>
  <label for="last_name">
    Last name: <span>*</span>
  </label>
  <input
    id="last_name"
    type="text"
    name="lastname"
    required
  />
  </div>
```

```
    </div>
    <div class="single-input">
      <label for="address">
        Address: <span>*</span>
      </label>
      <input
        id="address"
        type="text"
        name="address"
        required
      />
    </div>
```

3. Now, below the address field, we write the HTML and CSS for the postcode using `input` `id` and the country using `select`, as shown in the wireframe:

```
    <div class="double-input">
      <div>
        <label for="postcode">
          Postcode: <span>*</span>
        </label>
        <input
          id="postcode"
          type="text"
          name="postcode"
          required
        />
      </div>
      <div>
        <label for="country">Country:</label>
        <div class="select-wrapper">
          <select id="country">
            <option value="england">England</option>
            <option value="scotland">Scotland</option>
            <option value="ireland">Ireland</option>
            <option value="wales">Wales</option>
          </select>
        </div>
      </div>
    </div>
```

4. As shown in the preceding wireframe, we will now add a level 2 header followed by the `radio`-type checkboxes, as shown in the following code:

```
    <h2>Shipping method</h2>
    <div class="radio>
```

```
<input
  id="standard"
  type="radio"
  name="shipping-method"
  value="standard"
/>
<label for="standard">Standard</label>
</div>
<div class="radio">
  <input
    id="nextday"
    type="radio"
    name="shipping-method"
    value="nextday"
  />
  <label for="nextday">Next day</label>
</div>
<button type="submit">Submit</button>
```

5. If you look at the HTML file in your web browser, you should now have a web page that looks like the following screenshot:

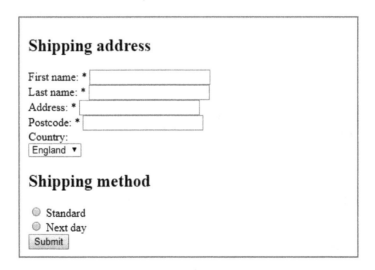

Figure 4.27: An unstyled checkout form in the browser

6. We can now add some CSS to style the form so that it more closely resembles the form shown in the preceding wireframe. We will start by applying some styles that deal with the spacing of elements and style the red asterisks shown in the wireframe:

```css
fieldset {
  border: 0;
  padding: 0;
}
fieldset > div {
  margin-bottom: 30px;
}
label {
  display: block;
  margin-bottom: 10px;
}
label span {
  color: red;
}
```

Note that `input` and `select` have the same style, as shown in the following code:

```css
input,
select {
  border: 1px solid grey;
  padding: 10px;
  width: 200px;
}
```

7. Now, let's add the CSS for the `select` boxes. We style `select` as shown in the following code snippet:

```css
select {
  background: transparent;
  border-radius: 0;
  box-shadow: none;
  color: #666;
  -webkit-appearance: none;
  width: 100%;
}
```

Complete styling `select` using the following code:

```css
.select-wrapper {
  position: relative;
  width: 222px;
}
```

```
.select-wrapper:after {
  content: '< >';
  color: #666;
  font-size: 14px;
  top: 8px;
  right: 0;
  transform: rotate(90deg);
  position: absolute;
  z-index: -1;
}
```

8. Finally, we will finish off the styling by adding some CSS for the inputs and the button styles, as shown in the following code:

```
.single-input input {
  width: 439px;
}

.double-input {
  display: flex;
}

.double-input > div {
  margin-right: 15px;
}

.checkbox input {
  float: left;
  width: auto;
  margin-right: 10px;
}
```

The following code is added to style the button:

```
button {
  background: green;
  border: 0;
  color: white;
  width: 224px;
  padding: 10px;
  text-transform: uppercase;
}
```

To style the inputs, we add the following code:

```
input:valid {
  border: 2px solid green;
}

input:invalid {
  border: 2px solid red;
}
```

If you now right-click on the filename in VS Code, on the left-hand side of the screen, and select **open in default browser**, you will see the form in your browser.

You should now have a form that looks like the following figure:

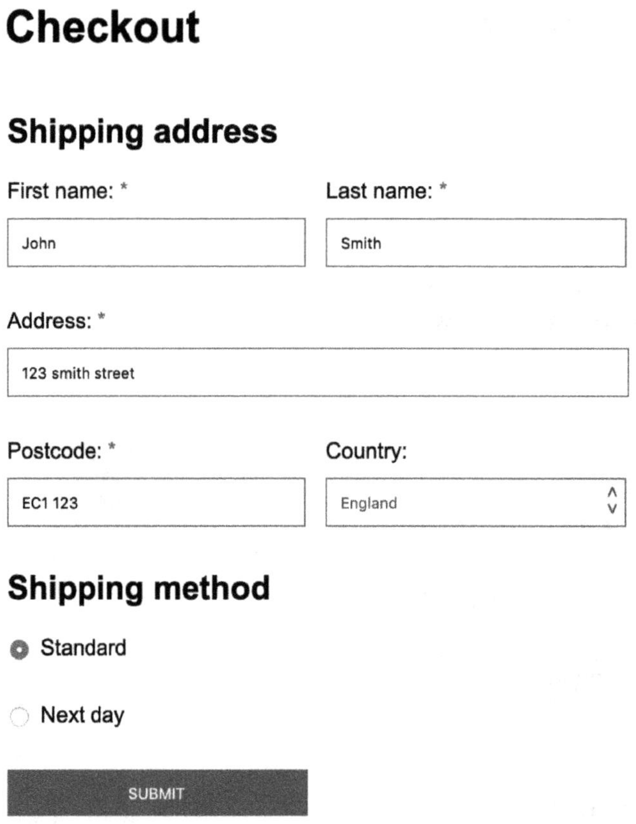

Checkout

Shipping address

First name: *

John

Last name: *

Smith

Address: *

123 smith street

Postcode: *

EC1 123

Country:

England

Shipping method

⊙ Standard

○ Next day

SUBMIT

Figure 4.28: A styled checkout form in the browser

9. Now, we will check whether the validation works for our required fields by submitting the form with empty fields. We will see something like the following screenshot:

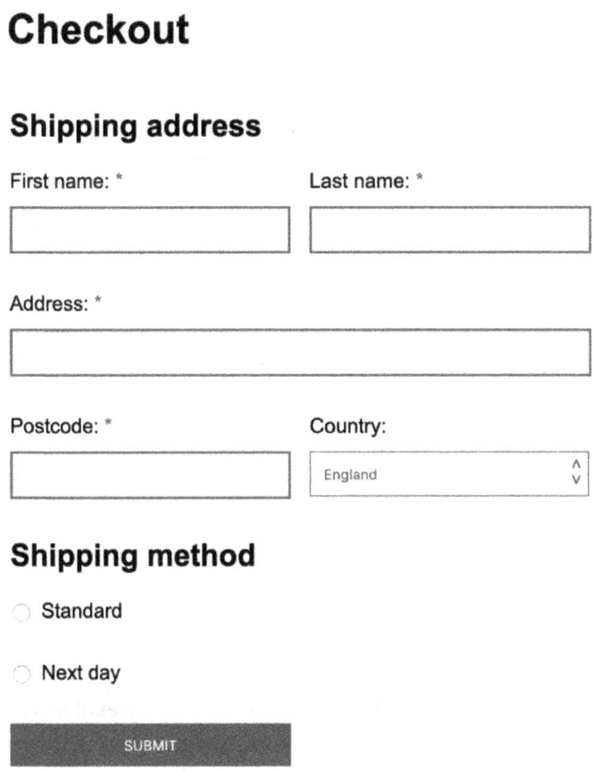

Figure 4.29: A form highlighting the required fields before submitting it

Activity 4.01 – building an online property portal website form

An international online property website has approached you to design a search form for their listings page. This form should have the following fields – `search radius`, `price range`, `bedrooms`, `property type`, and `added to the site` – as well as an option to include sold properties in the user's search. Create your own solution using the skills you have learned in this chapter:

1. Start by creating a new file named `form.html` in VS Code.

2. Then, start writing the HTML for the form using the description of the fields required for this form.

3. Once you are happy with the HTML you have written, you can let your creativity run wild and style the form using CSS. Make sure that you include styling for validation:

Property for sale in London

Search radius:	This area only
Price range:	Any
Bedrooms:	Any

Property type:	Any
Added to site:	Anytime
	☐ Include sold properties

FIND PROPERTIES

Figure 4.30: The expected output of the activity

> **Note**
> The solution to this activity is available on GitHub at `https://packt.link/7OoUW`

Summary

In this chapter, we continued our journey into building web pages by exploring web forms. We first studied the most common form-based HTML elements, including inputs, select boxes, textareas, and buttons. We then looked at the most common styling methods to style forms. To put this new knowledge into practice, we then built different forms.

We took some time to understand when you should use checkboxes and when to use radio buttons. We also spent some time looking at how you can add validation styles for web forms. We explored incorporating HTML tables for the layout of our forms and also looked at ways to make our forms more accessible.

In the next chapter of this book, we will learn how to take our web pages to the next level. We will learn how to make our web pages even more interesting by adding video and music to them.

5

Adding Animation to Web Pages

As we delve deeper into user experience research and UI design, we understand that a great application not only enables users to accomplish tasks but also guides them through the process seamlessly. A beautiful interface should feel intuitive to users, enhancing adherence and reducing the time spent on learning how to navigate it. Animations play a crucial role in achieving this goal. As developers, we can design user-friendly, natural, and fluid interfaces that facilitate interaction.

In this chapter, we'll explore the world of CSS animations to learn how to use them to enhance user experience in our applications, provide status feedback, and manage multiple complex animations. We're going to cover the following main topics:

- CSS animations using transitions
- CSS animation properties
- The `@starting-style` rule
- CSS positioning and z-index
- Complex animations with keyframes

Technical requirements

Coding in HTML/CSS can be done using any preferred IDE or text editor. For this chapter, **Visual Studio Code** (**VS Code**) will be used. You can use any browser you prefer, but I encourage you to use an up-to-date version of Google Chrome.

To view the coding results for this chapter and the CSS animation examples, please visit our GitHub repository to see the animations in action: `https://packt.link/gxGLm`

You can also clone or fork the data to experiment with the animations yourself—feel free to code along!

The transition property

The first step in CSS animation is understanding the transition property and its values. A transition is a property that enables CSS elements to change from one state to another in a smooth, animated manner.

Consider the following example:

```
.target {
  font-size: 14px;
  transition: font-size 4s 1s;
}

.target:hover {
  font-size: 36px;
}
```

In this example, the transition property is applied to the `target` class to change the `font-size` attribute from 14 px to 36 px on hover. Within this example, we can observe the following:

- The transition property is described within the element's original value, not in the changed one
- It's possible to specify the element that will change within the transition property
- We can set the transition duration immediately after describing which property will be the transition target or as the first attribute (**4s**)

Consider another example:

```
.target {
  font-size: 14px;
  color: red;
  transition: 4s ease-in-out 1s;
}

.target:hover {
  font-size: 36px;
  color: blue;
}
```

In the preceding example, we've added the `color` property to the `target` class. If we apply this CSS, we'll notice that both properties (`color` and `font-size`) are influenced by the transition:

- When the transition property is not specified, it applies to all other properties of the element.
- We can define `transition-timing-function` to specify the transition speed curve. In the preceding example, `ease-in-out` specifies a transition effect with a slow start and end.

- We can introduce a delay to the transition using `transition-delay` or its shorthand after the transition duration (`1s`).

Transition properties

The following table outlines the key transition properties, providing a brief description and the type of value each property accepts. These properties allow developers to control various aspects of the transition effect, such as the duration, timing, and specific CSS properties that should transition:

Property	Type of value	Description
transition (shorthand)	Set of values	A shorthand for combining the transition properties into a single property.
transition-property	CSS property	Determines the CSS property name for which the transition effect is applied.
transition-duration	Seconds or milliseconds	Specifies the duration, in seconds or milliseconds, for the transition effect to complete.
transition-timing-function	cubic-bezier(n,n,n,n) or ease | ease-in | ease-out | ease-in-out | linear | step-start | step-end	Defines the speed curve of the transition effect.
transition-delay	Seconds or milliseconds	Sets a delay, in seconds, before the transition effect begins.
[NEW!] transition-behavior	normal | allow-discrete	Specifies whether transitions will be started for properties whose animation behavior is discrete.

Table 5.1 – The transition properties and its accepted values

> **Important note**
>
> `transition-behavior` is a new property that doesn't work in older browsers by default. When using `{transition-behavior: allow-discrete}` to set discrete animations, make sure this is at the end of the CSS block to avoid specificity conflicts with the transition shorthand.

The transition shorthand follows a specific order of properties: *first, the property name; second, the duration; third, the timing function; fourth, the transition delay; and finally, the transition behavior.*

For instance, in the following shorthand, `opacity` is the property name, `2s` is the duration, `ease-in` is the timing function, `0.5s` is the delay, and `normal` is the transition behavior. If any of these properties are not specified, their default values will be used:

```
transition: opacity 2s ease-in 0.5s normal
```

Exercises with transitions

After understanding the theory behind CSS transitions, it's time to put that knowledge into practice. Let's create CSS animations that will provide a solid foundation for further improvement and can be applied to various projects. By working through these exercises, you'll gain hands-on experience that will enhance your ability to craft dynamic and engaging web animations.

Exercise #1 – simple menu animation

Our first exercise will be a simple, elegant animated menu. Begin by creating a project folder titled `CSS-animations-exercise-transitions`. Open this folder in VS Code and create two essential files: `index.html` and `style.css`.

> **Note**
> While it's possible to embed CSS styles directly within the HTML file using the `<style>` tag within `<head>`, it's advisable to separate responsibilities for a cleaner HTML file and to avoid unnecessary clutter.

Here's the initial HTML file:

```html
<!DOCTYPE html>
<html lang="en">
<head>
  <meta charset="UTF-8">
  <meta name="viewport" content="width=device-width,
        initial-scale=1.0">
  <title>CSS Animated Menu</title>
  <link rel="stylesheet" href="style.css">
</head>
<body>
  <aside>
    <ul>
      <li>
        <a href="#">
          <img src="./svg/new.svg"
               alt="New Document"
               class="menu-icon">
```

```
              <span>New Document</span>
          </a>
      </li>
      <li>
        <a href="#">
          <img src="./svg/recent.svg"
              alt="Open recent"
              class="menu-icon">
          <span>Recent</span>
        </a>
      </li>
      <li>
        <a href="#">
          <img src="./svg/search.svg"
              alt="Search documents"
              class="menu-icon">
          <span>Search</span>
        </a>
      </li>
    </ul>
  </aside>
</body>

</html>
```

In this file, we've crafted a straightforward menu list enclosed within the `<aside>` tag.

A single menu list is constructed using an unordered list (`ul`) comprising three list items (`li`). Each item incorporates an image element (`img`) and a text element (`span`), enveloped within an anchor (`a`) tag to denote links to other pages.

Image download instructions

To streamline the process, we've organized the images into a folder, ready for you to download. Please procure them from the provided link (`https://packt.link/shFG2`) and save them within a newly created folder named `svg`. Ensure this folder is nested within the project directory to maintain proper file paths.

Linking HTML and CSS files with <link>

Inspect the preceding HTML snippet. Within the `<head>` tag, we'll find the following element:

```
<link rel="stylesheet" href="style.css">
```

This tag establishes a connection between our HTML file and our CSS file. It's customary to name the primary CSS file `style.css` or `styles.css` and position it within the same folder as our HTML. With this link established, let's proceed to create the CSS file and commence styling!

Initial styling

Access the base CSS file and input the following code snippet. We'll dissect it line by line in sequential order:

```css
/* CSS RESET */
* {
  margin: 0;
  padding: 0;
  box-sizing: border-box;
}

/* INITIAL STYLES */

aside {
  width: 280px;
  background-color: #093c9b;
  padding: 10px;
  height: 100vh;
}

ul li {
  list-style: none;
  padding: 10px 15px;
  background-color: #053081;
  border-radius: 10px;
  margin: 10px;
}

ul li a {
  display: flex;
  gap: 20px;
  align-items: center;
  color: #fff;
  text-decoration: none;
  font-family: 'Verdana', sans-serif;
  font-size: 16px;
  line-height: 1.6;
}
```

```
ul li a img {
  width: 20px;
  height: 20px;
}
```

Understanding the file

The code starts with a **CSS reset**. This technique is frequently employed to eliminate any default CSS applied by the browser to HTML files. It standardizes the margin and padding of all elements to 0 and ensures that the box-sizing property is set to border-box for all elements, thereby computing the width inclusive of padding and border sizes. This approach establishes a uniform foundation for styling across various browsers:

```
* {
  margin: 0;
  padding: 0;
  box-sizing: border-box;
}
```

Styling the aside element

The <aside> element serves as the container for our menu list. The provided code configures its width, background color, padding, and height:

```
aside {
  width: 280px;
  background-color: #093c9b;
  padding: 10px;
  height: 100vh;
}
```

Note that the background color is specified using hexadecimal encoding, allowing for easy experimentation with different color values.

Styling the list items (ul li)

These styles target the <li> elements nested within <ul>. They eliminate the default list style and specify the padding, background color, border radius, and margin for each list item, enhancing their visual presentation:

```
ul li {
  list-style: none;
  padding: 10px 15px;
  background-color: #053081;
```

```
    border-radius: 10px;
    margin: 10px;
}
```

Styling anchor elements within list items (ul li a)

These styles are targeted at the anchor <a> elements nested within <li> elements. They employ
display: flex for layout, introduce spacing between elements, vertically align items, define the
text color, remove text decoration, and specify the font family, size, and line height, enhancing the
appearance and readability of the links:

```
ul li a {
    display: flex
    gap: 20px;
    align-items: center;
    color: #fff;
    text-decoration: none;
    font-family: 'Verdana', sans-serif;
    font-size: 16px;
    line-height: 1.6;
}
```

Styling images within anchor elements of list items (ul li a img)

These styles target <img> elements, which are descendants of <a> elements within <li> elements.
They establish the width and height of the images to be 20 px each, ensuring uniformity and appropriate
sizing within the list items:

```
ul li a img {
    width: 20px;
    height: 20px;
}
```

Enhancing CSS cohesion

To uphold the cascading nature of CSS, we adopt a structured approach to writing our styles, aligning
them with the HTML structure. We begin by styling the container, then progress to the list items,
and finally, address the inline elements, such as <img> and <a>. This is the **cascading rule** of CSS:

- **Specificity**: CSS selectors are prioritized based on specificity. More specific selectors override
  less specific ones. For instance, #id selectors hold more weight than .class selectors, which,
  in turn, supersede **element** selectors.

- **Source order**: In cases of conflicting styles with identical specificity, the style declaration that appears last in the CSS file, or is closest to the targeted element in an external stylesheet, takes precedence.

This hierarchical structure ensures the systematic application of styles and empowers developers to manage the visual presentation of HTML elements efficiently.

The initial result of our exercise will be as follows:

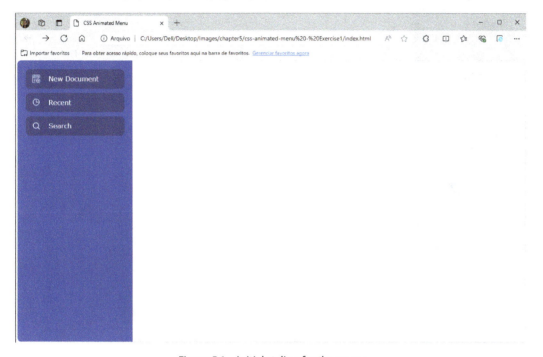

Figure 5.1 – Initial styling for the menu

Introducing our initial animation

Let's animate the menu list items by adjusting their background color when the mouse hovers over them. Initially, we define the style we want to apply when the element is in the :hover state:

```
ul li:hover {
    background-color: #0a4fbd;
}
```

This CSS declaration targets <li> elements within <ul> elements and alters their background color when they're hovered over by the cursor.

Upon hovering over the `<li>` element, we'll notice an immediate color change. However, this change lacks smoothness and natural transition, as it's not yet animated.

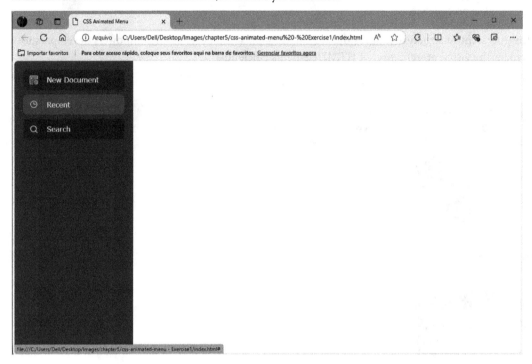

Figure 5.2 – Applying the :hover styling to the menu

Now, we're integrating the transition element into the primary `ul li` selector to dictate the animation characteristics. Add the following line within the `ul li` CSS properties:

```
transition: background-color 0.5s ease-in-out;
```

This line utilizes the shorthand transition property to encompass all the necessary transition properties for our animation. It's a concise alternative to individually specifying the transition properties, as demonstrated here:

```
transition-property: background-color;
transition-duration: 0.5s;
transition-timing-function: ease-in-out;
```

Here's a breakdown of the components:

- `transition-property: background-color` determines the CSS property that will smoothly transition. In this scenario, it indicates the transition of the background color of an element.

- `transition-duration: 0.5s` denotes the length of the transition effect. It's set to 0.5 seconds, indicating that any changes to the `background-color` property will transition over half a second:

  - Note: In JavaScript, it's advisable to use **milliseconds (ms)** for time measurements to maintain consistency

- `transition-timing-function: ease-in-out` specifies the timing function employed for the transition. This function governs the rate of change of the transition effect over time. `ease-in-out` initiates slowly, accelerates midway, and then decelerates toward the end, resulting in a smooth and natural transition effect.

Overall, this property declaration ensures that when the background color of an element changes, it does so smoothly over a duration of 0.5 seconds with a gradual acceleration and deceleration effect.

Congratulations! Our first animation is now complete. Take a moment to review the HTML file and observe the animation in action. Feel free to experiment with the transition further by adjusting parameters such as duration, delay, and other values to customize the effect.

Here's the final CSS file for this exercise:

```css
/* CSS RESET */
* {
  margin: 0;
  padding: 0;
  box-sizing: border-box;
}

/* INITIAL STYLES */

aside {
  width: 280px;
  background-color: #093c9b;
  padding: 10px;
  height: 100vh;
}

ul li {
  list-style: none;
  padding: 10px 15px;
  background-color: #053081;
  border-radius: 10px;
  margin: 10px;
  transition: background-color 0.5s ease-in-out;
}
```

```
ul li:hover {
  background-color: #0a4fbd;
}

ul li a {
  display: flex;
  gap: 20px;
  align-items: center;
  color: #fff;
  text-decoration: none;
  font-family: 'Verdana', sans-serif;
  font-size: 16px;
  line-height: 1.6;
}

ul li a img {
  width: 20px;
  height: 20px;
}
```

This code will produce this result:

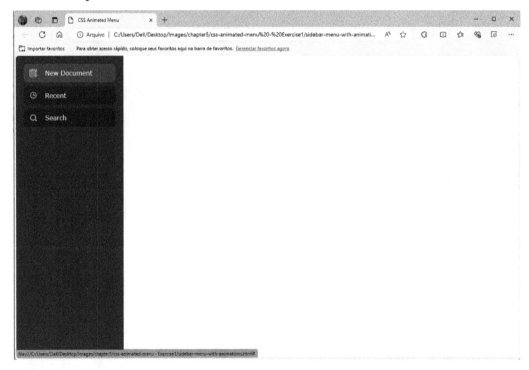

Figure 5.3 – Visual representation of the exercise coding

The following are common issues that might be encountered in this exercise:

- Is the CSS not being applied?

  - Ensure that the CSS file resides in the same directory as the HTML file

  - Double-check the placement and spelling of the (`<link rel="stylesheet" href="style.css">`) link tag within the HTML head tag

- Is the CSS functioning but some elements remain unstyled?

  - Verify the accuracy of class names and properties. Remember, CSS is case-sensitive.

  - Remember to save changes (*Ctrl + S*) and refresh the HTML file in `browser.qwq`.

If the code has been reviewed and appears identical to the instructions, consider downloading the provided source code to facilitate learning.

For detailed resolution, refer to the resolution code available at this link: `https://packt.link/g24bq`

Exercise #2 – adding more animations to the menu

The menu already boasts a polished style, but let's elevate it further. We'll enhance it by making the font weight bolder when hovering over the `<li>` element.

For this second exercise, duplicate the HTML file in the same folder and rename it to `sidebar-menu-with-animations.html`.

Inspect the `<li>` elements:

```
<li>
        <a href="#">
          <img src="./svg/search.svg"
              alt="Search documents"
              class="menu-icon">
          <span>Search</span>
        </a>
      </li>
```

Each `<li>` contains an `<img>` element with the menu-icon class assigned and a `<span>` tag containing the link label. We'll animate the `<img>` tag to enlargen when hovered over and make the text bolder simultaneously:

```
ul li:hover a {
    font-weight: bold;
}

ul li:hover a img {
    width: 24px;
    height: 24px;
}
```

Interpreting this CSS code, the browser is instructed that when hovering over the `<li>` element, the child `<a>` element will have a bolder font weight, and the child `<img>` element will increase in size. Even when altering the behavior of a child element, the trigger (hover) can be set on the parent element to execute the specified behavior.

Upon reviewing the application, it becomes apparent that while the transition of the background persists, the newly defined properties aren't transitioning. This occurs because although we can set the :hover trigger to a parent in CSS, we must still specify the transition values into the element itself. Let's rectify this:

```
ul li a {
    display: flex;
    gap: 20px;
    align-items: center;
    color: #fff;
    text-decoration: none;
    font-family: 'Verdana', sans-serif;
    font-size: 16px;
    line-height: 1.6;

    transition: font-weight 0.5s;
}

ul li a img {
    width: 20px;
    height: 20px;
    transition: all 0.5s;
}
```

Enhancing animation performance – key points

In the menu's `<a>` tag, there are several CSS properties defined, whereas only two properties are specified for the image in the `<img>` tag.

To enhance animation efficiency, we specify which property will be animated within the `<a>` element. This optimization is crucial because the transition property defaults to animating all properties, which can potentially slow down the application.

However, in the `<img>` tag, where only the desired properties are targeted for animation, we can either use the `all` value or omit it altogether, as it defaults to transitioning all properties. This selective approach streamlines the animation process and improves overall performance.

Congratulations! Our animated sidebar menu is now complete!

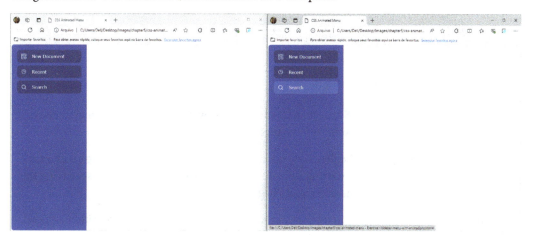

Figure 5.4 – Animation effect applied in the menu

The CSS file should be looking like this:

```
/* previous CSS */

ul li a {
  display: flex;
  gap: 20px;
  align-items: center;
  color: #fff;
  text-decoration: none;
  font-family: 'Verdana', sans-serif;
  font-size: 16px;
  line-height: 1.6;
  transition: font-weight 0.5s;
```

```
}

ul li:hover a {
   font-weight: bold;
}

ul li a img {
   width: 20px;
   height: 20px;
   transition: 0.5s;
}

ul li:hover a img {
   width: 24px;
   height: 24px;
}
```

The @starting-style rule

To prevent unexpected outcomes, CSS transitions do not activate by default when an element undergoes its initial style update or when its display type changes from none to another value.

For these transitions to occur on the first style update, it's possible to use the new @starting-style rules. These rules define the starting styles for elements that lack a prior state, specifying the property values from which the transition should begin.

You can utilize @starting-style in two ways: either as a standalone rule or nested within a ruleset.

The standalone rule looks as follows:

```
@starting-style {
   selector {
      properties
   }
}
```

The nested rule looks as follows:

```
selector {
   properties

   @starting-style {
      properties
   }
}
```

@starting-style proves particularly beneficial for creating entry and exit transitions for elements that appear in the forefront (such as popovers and modal dialogs), elements transitioning to and from display: none, and elements newly added to or removed from the DOM. Let's see how it works in the next exercise.

Exercise #3 – multiple transitions with the @starting-style rule

In this project, we aim to implement an entry animation for each <li> item in the menu of *Exercise #2*. The items will smoothly slide in from the left upon appearance. To achieve the desired behavior, we will utilize a combination of named transitions and the @starting-style rule.

Let's create a multiple-transition effect in the *Exercise #2* menu:

1. Open the folder for *Exercise #2*. No changes are required in the HTML file; our focus will be on the CSS.

2. Begin by defining the initial style for the <li> elements. They should be invisible and positioned outside the viewbox to allow for a sliding effect. Add the following code to the end of the CSS file:

```css
@starting-style {

  ul li {
    opacity: 0;
    transform: translateX(-280px);
  }

}
```

3. Define the final (default) state for the <li> elements, incorporating the properties specified in the starting style to enable transitions:

```css
ul li {
  list-style: none;
  padding: 10px 15px;
  background-color: #053081;
  border-radius: 10px;
  margin: 10px;
  transition:
    background-color 0.5s ease-in-out;
  opacity: 1;
  transform: translateX(0px);
}
```

> **Note**
>
> Setting `translateX` to 0 restores the element to its default position.

4. Introduce the animation by adding the `opacity` and `transform` properties to the transition:

```css
ul li {
list-style: none;
padding: 10px 15px;
background-color: #053081;
border-radius: 10px;
margin: 10px;
transition:
background-color 0.5s ease-in-out,
opacity 0.5s ease-in-out,
transform .7s ease-in-out;
opacity: 1;
transform: translateX(0px);
}
```

> **Note**
>
> Specifying the properties we want to animate *enhances performance* by targeting only those properties that change.

The result of our code is as follows:

Figure 5.5 – Animated transition effect

While the animation result is satisfactory, all `<li>` elements animate simultaneously, and we desire a cascading effect. To achieve this, introduce a transition delay for each element:

```css
ul li:nth-child(1) {
   transition-delay: 0.1s;
}
ul li:nth-child(2) {
```

```
    transition-delay: 0.2s;
}
ul li:nth-child(3) {
    transition-delay: 0.3s;
}
```

Congratulations! Our entry animation is now complete. Enjoy the stunning result!

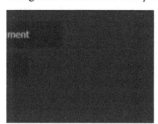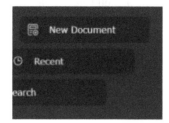

Figure 5.6 – Representation of the developed animation flow

The code containing the result of this exercise is accessible from the link provided here: https://packt.link/j12MO

> **Note**
>
> @starting-style is only relevant to CSS transitions and doesn't affect animations with keyframes. The @starting-style rule is a new feature for CSS. At the time of writing this book, it's only available to the browsers in this list: https://caniuse.com/mdn-css_at-rules_starting-style.

After mastering the basics of CSS transitions, it's time to explore more complex techniques essential for creating dynamic effects. In the following section, we'll learn about advanced concepts such as **positioning**, **z-index**, **blur**, and **opacity**. These skills will enhance your ability to craft sophisticated and engaging animations for your web projects.

Advanced CSS for animations

The transition property facilitates common animation effects; however, it is crucial to explore additional resources to enhance user experience and the perceived value of our application.

To craft more intricate animations, it's beneficial to acquaint ourselves with complementary properties that, when integrated, can produce compelling and impressive CSS animations.

CSS positioning

The `position` property specifies how an element should be positioned within its containing element. There are five possible values for the `position` property: `static`, `relative`, `fixed`, `absolute`, and `sticky`. Each value determines how the element interacts with its surroundings and how it is positioned on the page:

- `position: static`: This is the default value. Elements with `position: static` are positioned according to the normal flow of the document. They are not affected by the `top`, `bottom`, `left`, and `right` properties.

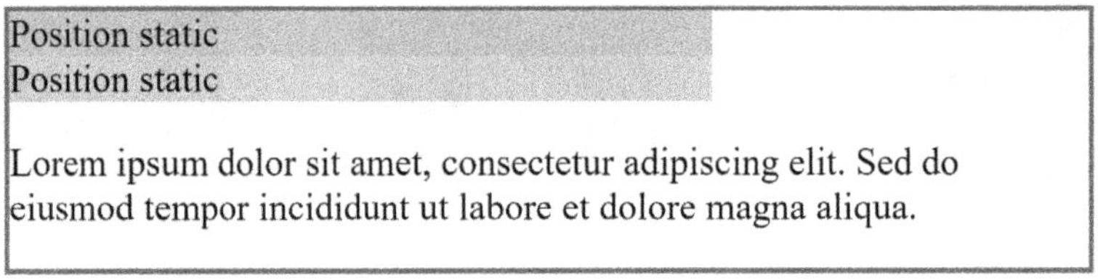

Figure 5.7 – position: static representation

- `position: relative`: Elements with `position: relative` are positioned relative to their normal position in the document flow. This property allows us to use the `top`, `bottom`, `left`, and `right` properties to move them from their original position.

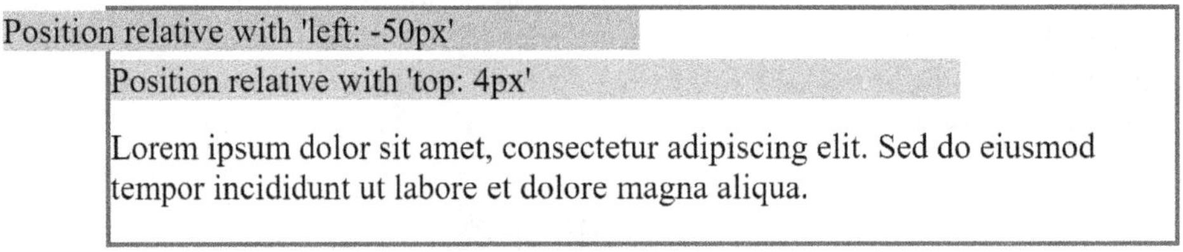

Figure 5.8 – position: relative representation

- `position: fixed`: Elements with `position: fixed` are positioned relative to the viewport, meaning they remain in the same place even when the page is scrolled. We can use the `top`, `bottom`, `left`, and `right` properties to specify their position.

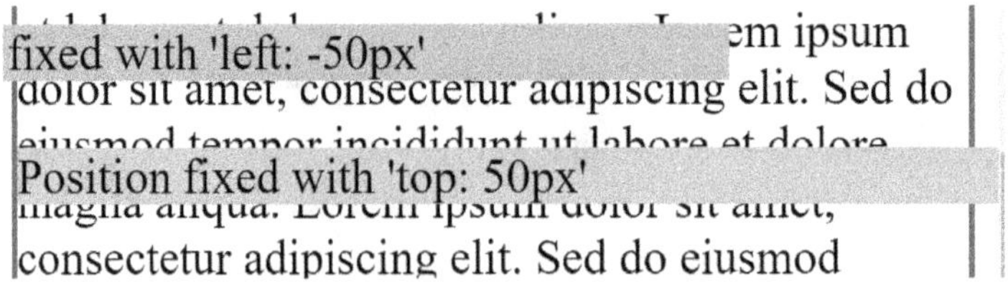

Figure 5.9 – position: fixed representation

- `position: absolute`: Elements with `position: absolute` are positioned relative to the nearest positioned ancestor. If no positioned ancestor is found, they are positioned relative to the initial containing block, usually the `<html>` element. They are removed from the normal document flow and can overlap with other elements, as if they are in a layer on top of the original document.

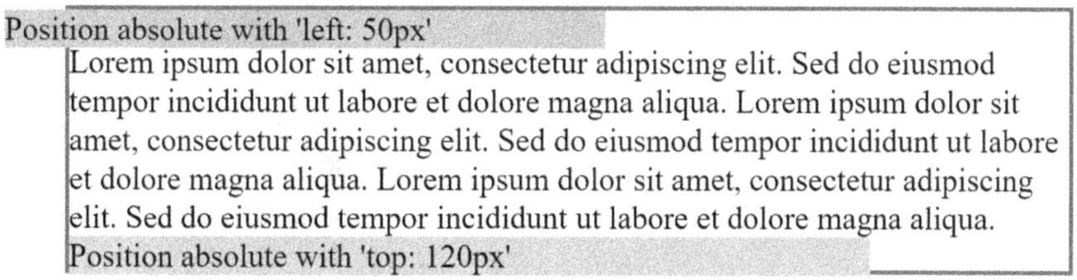

> Position absolute with 'left: 50px'
>
> Lorem ipsum dolor sit amet, consectetur adipiscing elit. Sed do eiusmod tempor incididunt ut labore et dolore magna aliqua. Lorem ipsum dolor sit amet, consectetur adipiscing elit. Sed do eiusmod tempor incididunt ut labore et dolore magna aliqua. Lorem ipsum dolor sit amet, consectetur adipiscing elit. Sed do eiusmod tempor incididunt ut labore et dolore magna aliqua.
> Position absolute with 'top: 120px'

Figure 5.10 – position: absolute representation

The `position` property in CSS enables us to modify the positioning of elements within our application.

We can adjust their placement based on the layout of their parent elements or in response to scrolling events. This capability provides us with a valuable tool for crafting intricate animations and achieving more dynamic and engaging user experiences.

z-index

When we assign the `position` property to an element, it enables us to utilize the `z-index` property. The `z-index` property governs the stacking order of an element, deciding whether it should be positioned in front of or behind other elements.

Its value is a numerical integer, which can be positive (higher layers) or negative (lower layers), with `0` often serving as the reference point within the application code.

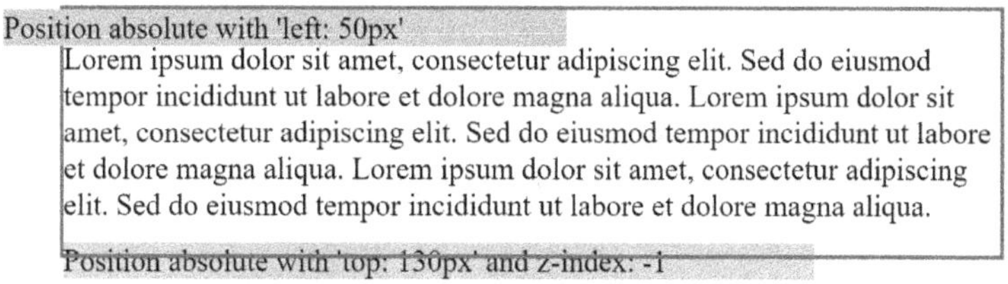

> Position absolute with 'left: 50px'
>
> Lorem ipsum dolor sit amet, consectetur adipiscing elit. Sed do eiusmod tempor incididunt ut labore et dolore magna aliqua. Lorem ipsum dolor sit amet, consectetur adipiscing elit. Sed do eiusmod tempor incididunt ut labore et dolore magna aliqua. Lorem ipsum dolor sit amet, consectetur adipiscing elit. Sed do eiusmod tempor incididunt ut labore et dolore magna aliqua.
>
> Position absolute with 'top: 130px' and z-index: -1

Figure 5.11 – z-index representation

Understanding `z-index` is important for animating with CSS because it allows us to control the stacking order of elements. This is crucial for managing overlapping effects, creating depth perception, and ensuring interactive elements remain visible during animations, which all contribute to a more polished and engaging user experience.

Opacity

CSS opacity allows us to control the transparency levels of elements. This property operates on a scale ranging from 0.0 (completely invisible) to 1 (fully visible), enabling precise adjustments to the element's transparency.

Opacity: 0.1   Opacity: 0.5   Opacity: 1

Figure 5.12 – The opacity property representation – photo by Fahmi Fakhrudin in Unsplash

The following CSS is applied to Figure 5.12:

```css
.image-container {
  margin-right: 20px;
}
.image-container:nth-child(1) img {
  opacity: 0.1;
}
.image-container:nth-child(2) img {
  opacity: 0.5;
}
.image-container:nth-child(3) img {
  opacity: 1;
}
```

Opacity is an important property for creating various types of animations within CSS. It allows us to control the transparency of an element, which can be used to create effects such as fading in or out, transitions between different states, and overlay effects.

Blur

The blur property in CSS enables us to apply a blur effect to elements, enriching visual design possibilities and facilitating creative effects in web development. It functions as a CSS filter, allowing us to have precise control over the degree of blurriness applied to targeted elements.

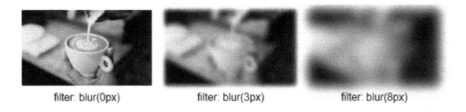

filter: blur(0px)   filter: blur(3px)   filter: blur(8px)

Figure 5.13 – The blur property representation – photo by Fahmi Fakhrudin in Unsplash

The following CSS is applied to Figure 5.13:

```
.image-container:nth-child(1) img {
  filter: blur(0px);
}
.image-container:nth-child(2) img {
  filter: blur(3px);
}
.image-container:nth-child(3) img {
  filter: blur(8px);
}
```

While there are a lot of CSS properties that can be combined to craft compelling animations, the ones mentioned previously are among the most commonly used. They can be integrated within the `transition` property for state animations or within the `@keyframes` CSS animation for more intricate, looping animations. This is what we'll delve into next.

The keyframes rule and CSS animation properties

Occasionally, enhancing user experience throughout their journey with our app necessitates incorporating more intricate animations. Therefore, let's delve deeper into understanding animation properties and the `@keyframes` rule.

To create a CSS animation sequence, we style the element we want to animate with the animation property or its sub-properties. This allows us to configure the timing, duration, and other details of how the animation sequence should progress.

However, it's important to note that this setup doesn't directly influence the visual appearance of the animation, as that aspect is controlled by the `@keyframes` rule.

The following are CSS animation properties and their shorthands:

Property	Values accepted	Description
animation (shorthand)	Set of values	A shorthand for combining the transition properties into a single property.
animation-delay	Seconds or milliseconds	
animation-direction	`normal`, `reverse`, `alternate` or `alternate-reverse`	Determines whether an animation should proceed in a forward direction, a backward direction, or alternate between playing the sequence forward and backward.
animation-duration	Seconds or milliseconds	Sets the length of time for an animation.
animation-fill-mode	`none`, `forwards`, `backwards`, `both`	Defines how a CSS animation applies styles to its target before and/or after its execution.
animation-iteration-count	`infinite` or number (default: 1)	The number of times an animation should play before stopping.
animation-name	keyframe name	Specifies the names of one or more `@keyframes` animations to be applied to an element
animation-play-state	`running` or `paused`	Sets whether an animation is running or paused.
animation-timeline	`none`, `auto`, `scroll()`, `view()` or `named`	Specifies the timeline that is used to control the progress of a CSS animation. This timeline can be the default document timeline, a scroll-triggered timeline, a view-triggered timeline, or even a custom-named timeline.
animation-timing-function	`cubic-bezier(n,n,n,n)`, `ease`, `ease-in`, `ease-out`, `ease-in-out`, `linear`, `step-start` or `step-end`	Determines the progression of an animation throughout each cycle's duration.

Table 5.2 – The animation properties and its accepted values

The shorthand for CSS @keyframes animation includes properties in a specific order:

1. **Duration**: Time taken for one animation cycle

2. **Easing function**: Rate of change over time

3. **Delay**: Time before animation starts

4. **Iteration count**: Number of animation cycles

5. **Direction**: Play direction (forward, backward, and alternate)

6. **Fill mode**: Styling before/after animation

7. **Play state**: Whether animation is running/paused

8. **Name**: Identifier for the animation

Default values apply if any property is omitted, except for duration and name, which are required for the animation to function. Here's an example of a @keyframes shorthand:

```
animation: 3s ease-in 1s 2 reverse both paused slidein;
```

The @keyframes rule dictates the visual behavior of an animated element at specific intervals throughout the animation sequence. In CSS, animations are timed through style configurations, with keyframes using percentages to denote their position within the sequence. At 0%, the animation begins, and at 100%, it reaches its conclusion.

These key moments can be referred to as from and to, respectively, and are optional. If these points are not explicitly defined, the browser defaults to using the computed attribute values for animation initiation and completion. An example of keyframes to animate the filter: blur property is as follows:

```
@keyframes blurAnimation {
  from {
    filter: blur(0px); /* No blur at the beginning */
  }
  to {
    filter: blur(10px); /* Blurred effect at the end */
  }
}
```

Using the preceding example, we're applying the animation named blurAnimation to a div, using the CSS animation shorthand:

```
div {
  width: 200px;
  height: 200px;
  background-color: lightblue;
```

```
animation: blurAnimation 2s ease-in-out
        infinite alternate;
}
```

Initial State   Final State

Figure 5.14 – Keyframe blurAnimation applied to a div

Exercise #4 – loading spinner with keyframes

Let's create a div featuring a loading spinner to indicate to our users that their request is currently being processed. This is a straightforward and frequently employed application of CSS animations, which, depending on the architecture of the application, minimizes performance degradation compared to .gif files or JavaScript animations.

Before we begin the exercise, please download the image asset from the following link: https://packt.link/ypVhx. Once downloaded, ensure to place the spinner.svg file in the same folder as the HTML file.

The HTML code will include a div to contain the spinner and an image element for the spinner itself:

```
<!DOCTYPE html>
<html lang="en">
<head>
  <meta charset="UTF-8">
  <meta name="viewport" content="width=device-width,
        initial-scale=1.0">
  <title>Load Spinner</title>
  <link rel="stylesheet" href="style.css">
</head>
<body>
  <div class="container">
    <img src="./spinner.svg" alt="">
  </div>
</body>
</html>
```

Now, let's create a file named `style.css` in the same folder as the HTML file. The initial styles for the elements are as follows:

```
* {
  margin: 0;
  padding: 0;
  box-sizing: border-box;
}

.container {
  display: flex;
  justify-content: center;
  align-items: center;
  height: 100vh;
}

img {
  width: 50px;
  height: 50px;
}
```

When you open the HTML file in your browser, you'll see a static image displayed:

Figure 5.15 – Static load spinner image

Now, let's create the animation using keyframes. The `transform` property is utilized to alter the appearance or position of a block element. In the upcoming keyframe, we'll employ `transform: rotate(360deg)` to make the image spin in a full circle and ensure its animation concludes at the same point where it started:

```
@keyframes loadSpinner {
  0% {
    transform: rotate(0deg);
  }
  100% {
    transform: rotate(360deg);
  }
}
```

Finally, we apply the `loadSpinner` animation to the `img` tag in CSS:

```
img {
  width: 50px;
  height: 50px;
  animation: loadSpinner 1s linear infinite;
}
```

Figure 5.16 – Visual representation of the spinner animation

The code containing the resolution for this exercise is accessible in the link provided here: `https://packt.link/tbpEJ`

Exercise #5 – menu icon shaking effect with keyframes

Let's enhance the user experience by adding a dynamic shaking effect to the menu icons from *Exercise #2*. Navigate to the directory of the *Exercise #2* project and follow these steps to implement the animation:

1. **Create keyframes**: Define a keyframe animation named **shake** to bring life to the menu icons. This animation will add a subtle yet engaging shaking motion to the icons when hovered over:

    ```
    @keyframes shake {
      0% { transform: translateX(0); }
      25% { transform: translateX(-5px) rotate(5deg); }
      50% { transform: translateX(5px) rotate(-5deg); }
      75% { transform: translateX(-5px) rotate(5deg); }
      100% { transform: translateX(0); }
    }
    ```

2. **Apply animation**: Assign the **shake** animation to the menu icons' hover state. This ensures that the animation activates whenever a user interacts with the icons:

    ```
    ul li:hover a img {
      width: 24px;
      height: 24px;
      animation: shake 0.4s ease;
    }
    ```

By following these steps, you'll add an interactive touch to the menu icons, making the user experience more engaging and visually appealing:

Figure 5.17 – Visual representation of the menu animation flow

The code containing the resolution for this exercise is accessible in the link provided here: `https://packt.link/36avH`

Summary

Great job! In this chapter, we delved deep into the realm of CSS animations. By now, you've undoubtedly honed your skills in crafting stunning, polished animations that elevate your application while minimizing performance overhead.

As you reflect on what you've learned about transitions, positioning, effects, and keyframes, remember that mastery comes with practice and exploration. Continuously immerse yourself in the vast landscape of CSS, as familiarity breeds proficiency.

I encourage you to select three or four HTML files from your projects and imbue them with precise, elegant animations to enhance the user experience and elevate the perceived value. Keep pushing the boundaries of your creativity, and watch as your skills flourish with each animation you create.

In the upcoming chapter, we'll explore CSS theming and color theory to craft applications with significant visual impact while prioritizing user preferences and ease of maintenance. Additionally, we'll dive into optimizing the user experience by implementing adaptable color themes that seamlessly adjust to the user's operating system or preferences.

6

Themes, Color, and Polishing Techniques

As web developers, we often face the challenge of accommodating every user's browser configuration to ensure the optimal experience with our application. Users may have preferences for larger font sizes, different text spacing, or even processing information in a specific color scheme. Whether due to accessibility needs, personal preferences, or adjusting for varying reading conditions, there are numerous variables that can impact how our code is interpreted.

To address these concerns, it's crucial to anticipate these variables by providing at least two different themes: one for users who prefer a dark mode and another for those who prefer a light mode in their browser settings. This practice not only enhances the **user experience (UX)** but also adds perceived value to our application. Moreover, it enables us to craft more tailored styles for different scenarios, ensuring consistency and usability across various environments.

In this chapter, we'll learn about the following topics:

- How CSS themes work
- CSS color variables
- Learning color theory with HSL
- Understanding the `invert()` filter
- Exploring user's preferences with `color-scheme`
- Introducing the `light-dark()` function
- Understanding typography and font choices

Let's delve into how themes function in CSS and explore the process of creating a new dark theme for our application!

Technical requirements

The code files for this chapter can be found at `https://packt.link/qTKYl`.

Introduction to web design themes

Color themes in web design work by defining a set of colors that are consistently applied throughout a website or application. These themes help to create a cohesive and visually appealing experience for users.

The first step in creating a color theme is to select a palette of colors that harmonize well together. This palette usually consists of a primary color, secondary colors, and accent colors. Designers often use tools such as Adobe Color, Coolors, or Color Hunt to create or find suitable color palettes. *Figure 6.1* represents a color palette:

Figure 6.1 – Color palette from the color-hex website (https://www.color-hex.com/color-palette/389)

Once the color palette is established, designers apply these colors consistently throughout the website or application. This includes text, backgrounds, borders, buttons, links, and other **user interface** (**UI**) elements. Consistency in color usage helps create a unified and coherent visual identity.

Many websites and applications offer both light and dark themes to accommodate user preferences and varying lighting conditions. Light themes typically feature lighter background colors with darker text, while dark themes invert this color scheme, using darker backgrounds with lighter text. Switching between light and dark themes can enhance readability and reduce eye strain, particularly in low-light environments. *Figure 6.2* shows a bookstore website rendered in light mode, as an example:

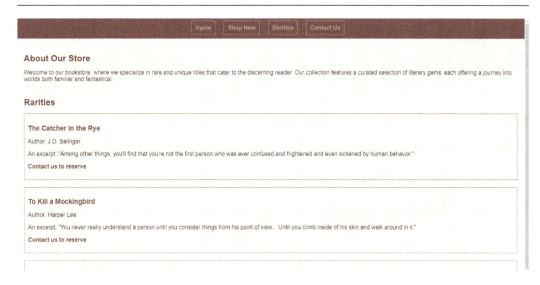

Figure 6.2 – Website styles with the cappuccino palette in light mode

Now, the same website is rendered in dark mode, as we can see in *Figure 6.3*:

Figure 6.3 – The same website, now in dark mode

Switching between light and dark themes using JavaScript has been a common practice. However, with the introduction of the `color-scheme` CSS property, we now have the capability to define styles based on the user's preferred theme directly in CSS, eliminating the need for JavaScript altogether. We'll explore this concept further in the upcoming exercises. Before we delve into that, let's first familiarize ourselves with the fundamental concept of variables in CSS.

CSS color variables

Variables in CSS, also known as **CSS custom properties**, allow you to define reusable values that can be used throughout your stylesheets. They provide a way to store values such as colors, sizes, fonts, or any other property value, making it easier to maintain and update styles across your entire project. Let's look at the following code:

```
:root {
  --primary-color:   #854442;
  --secondary-color: #4b3832;
  --text-color: #3c2f2f;
  --background-color: #fff4e6;
  --highlight-color: #be9b7b;
}
```

In this example, :root is a pseudo-class selector that targets the root element of the document, typically the <html> element. --primary-color is the name of the variable, preceded by two dashes (--). #854442 is the value assigned to the variable. Let's look at another piece of code:

```
header {
    background-color: var(--primary-color);
    color: var(--secondary-color);
    padding: 20px;
    text-align: center;
}
```

var(--primary-color) is how you use the variable. The var() function retrieves the value stored in the variable.

You can then update the value of the variable in one place, and it will automatically apply to all elements that use that variable. This makes it easy to make global style changes across your project.

CSS variables offer a powerful way to manage and customize styles, improve consistency, and streamline the development process. They are supported in modern browsers and can be used alongside other CSS features to create dynamic and flexible designs.

Exercise 1 – creating a light theme using CSS color variables

In this exercise, we'll be enhancing the appearance of the bookstore website introduced earlier. We'll begin with the raw HTML structure and a basic CSS file that lacks any coloring or styling. To proceed, please download the initial assets using the following link: https://packt.link/uLZRc

1. Open the folder containing the project files using VS Code. Create a new file named `light.css` within the same directory as your existing `style.css` and `bookstore.html` files. This `light.css` file will host all the color variables for our application in light mode. Populate the file with the following variable declarations:

```
:root {
    --primary-color:    #0c457d;
    --secondary-color: #e8702a;
    --text-color: #061b1a;
    --background-color: #fcf6ed;
    --highlight-color: #6bd2db;
}
```

2. Next, we'll link these colors to our `bookstore.html` file and `style.css`. Insert a new `<link>` element within the `<head>` section of our HTML file to reference the `light.css` file:

```
<html lang="en">
<head>
    <meta charset="UTF-8">
    <meta
        name="viewport"
        content="width=device-width,
                initial-scale=1.0"
    >
    <title>Rarities - Bookstore</title>
    <link rel="stylesheet" href="style.css">
    <link rel="stylesheet" href="light.css">
</head>
<body>
```

Be sure not to replace any existing elements; simply add the second `<link>` element alongside the first one. This ensures that both CSS files are linked to the HTML document, allowing us to seamlessly integrate the defined color variables into our styles.

3. Now that we've defined our color variables in the HTML file, let's integrate them into our main `style.css` file. Open `style.css` in VS Code and add the following declarations within the body selector to apply the color variables:

```
body {
    font-family: Arial, sans-serif;
    margin: 0;
    padding: 0;
    color: var(--text-color);
    background-color: var(--background-color);
}
```

4. Let's enhance the header by adding color to its background. Notice that the name of the variable doesn't limit its usage; it's possible to assign the color variable to any element of your choosing:

```css
header {
    padding: 20px;
    text-align: center;
    background-color: var(--primary-color);
}
```

5. Apply color to the anchor elements (`<a>`) inside the header:

```css
header nav ul li a {
    text-decoration: none;
    color: var(--highlight-color);
}
```

6. Add color to the text elements within the headers (`<h1>`, `<h2>`, `<h3>`):

```css
h1, h2, h3 {
    margin-bottom: 10px;
    color: var(--secondary-color);
}
```

7. Let's give each `<li>` element a colored border:

```css
li {
    margin-bottom: 20px;
    border: 1px solid var(--highlight-color);
    padding: 10px;
}
```

8. Lastly, style the anchor elements (`<a>`) inside the book cards:

```css
a {
    font-weight: bold;
    text-decoration: none;
    color: var(--primary-color);
}
```

Congratulations! We've successfully styled our first themed application. The result should resemble *Figure 6.4*:

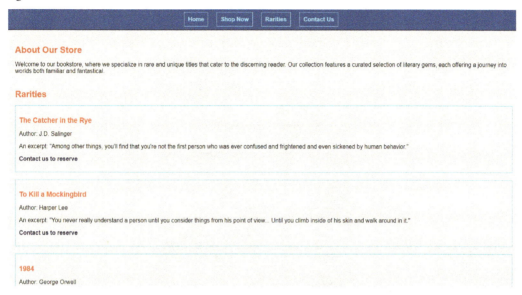

Figure 6.4 – Result of the application with color variables

Note

Using variables to define colors instead of directly applying them to individual elements offers several benefits.

Consistency: By defining colors as variables, you ensure consistency throughout your application. If you need to change a color later on, you only need to update the variable value once, and the change will be applied across all elements that use that variable.

Ease of maintenance: Managing colors centrally in variables makes your code base easier to maintain. You can easily locate and update colors without having to search through each occurrence of a specific color used in your stylesheets.

Reusability: Variables allow you to reuse colors across different elements and components without duplicating code. This promotes a more modular and scalable approach to styling your application.

Clarity and readability: Using variables makes your CSS code more readable and understandable. Instead of hardcoded color values, variables provide meaningful names that describe the purpose or role of each color, improving code clarity and developer comprehension.

Flexibility: Variables provide flexibility in styling. For instance, you can define different sets of colors for different themes (e.g., light mode, dark mode) and switch between them dynamically by updating the variable values, offering a more dynamic and customizable UX.

Learning color theory with HSL

CSS accepts color values in several formats, providing flexibility in specifying colors according to different needs and preferences. The most common usage is in hexadecimal notation, which represents the red, green, blue (RGB) color model. Hexadecimal notation starts with a # followed by six characters, where each pair represents the intensity of red, green, and blue respectively.

Figure 6.5 represents the colors in hexadecimal code from the Exercise 1:

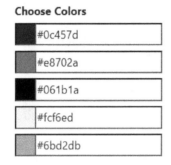

Figure 6.5 – Hexadecimal code of the used colors

Another way to write good CSS color values, especially when in need of creating different themes with opposite values, is to use the hsl() function.

In CSS, hsl() is a color function that stands for hue, saturation, and lightness. It allows you to define a color by specifying its **hue**, **saturation**, and **lightness** values.

Hue represents the type of color, expressed as an angle between 0 and 360 degrees on the color wheel. Red is at 0 degrees, green at 120 degrees, and blue at 240 degrees.

Saturation represents the intensity or purity of the color, expressed as a percentage. 0% saturation yields grayscale, while 100% saturation is the most vivid form of the color.

Lightness represents the brightness of the color, also expressed as a percentage. 0% lightness yields black, 100% lightness yields white, and 50% lightness is normal.

hsl() provides a flexible and intuitive way to manipulate colors in CSS, making it an excellent choice for creating dark themes or any other color variations where precise control over brightness is needed. The syntax is as follows:

```
Selector {
  color: hsl(0, 100%, 50%);
}
```

Lets understand more about colors in sequence.

The HSL color wheel

The crucial concept to understand about hue is its specification using an angle or degree value. It's common to see hues represented in a color wheel, like *Figure 6.6* shows:

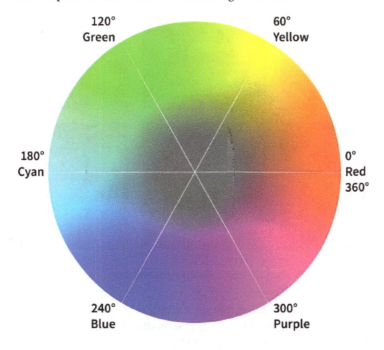

Figure 6.6 – Color wheel

Visualizing the color wheel offers a straightforward method of identifying complementary (or opposite) colors, such as red and cyan or green and purple.

When aiming for a color between two primary hues, selecting a number becomes intuitive. For instance, if I seek a shade of pink, I'd likely choose a number between 300 and 360 degrees.

After understanding hue, the concepts of saturation and lightness become clearer. Let's explore the following colors and their coding representation in *Figure 6.7*:

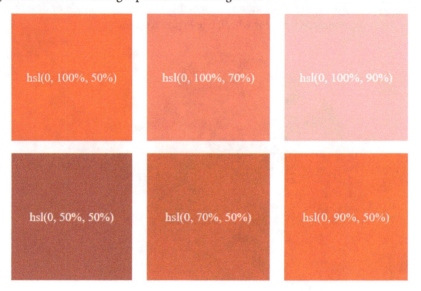

Figure 6.7 – HSL coloring representation

When choosing colors for your application, you must consider the **emotional impact** they convey, their **compatibility** with your brand identity, and their **accessibility** for all users. Remember to follow UI/UX polishing techniques to enhance the overall UX. In sequence, we'll see more polishing tips for your projects.

UX/UI polishing tips

Polishing techniques for the UI/UX involve refining and enhancing the visual and interactive aspects of a UI to create a more seamless and engaging UX. Here are some tips to effectively polish your design:

- Ensure visual consistency by maintaining a cohesive color scheme, typography, and iconography throughout the interface.

- Incorporate visual feedback, such as animations or transitions, to provide users with a more responsive and intuitive experience.

- Prioritize content hierarchy by organizing information in a logical and intuitive manner.

- Utilize visual cues, such as color, size, or typography, to emphasize important content and direct user attention.

- Ensure accessibility by following guidelines such as **Web Content Accessibility Guidelines (WCAG)** to make the interface usable for users with disabilities.

These details distinguish an ordinary website from a well-designed, engaging application crafted to maintain user interest for extended periods. Now, let's put this into practice in the next exercise.

Exercise 2 – creating a dark theme using hsl()

Now that we've gained insight into using hsl() colors and discovering their complementary values, let's apply this concept to create a dark theme for our bookstore from Exercise 1.

Please navigate to the project folder in VS Code and review the color variables we've defined in light. css. Since they're currently written in hexadecimal values, our initial step is to convert them to the hsl() format.

Fortunately, we can rely on online tools such as https://htmlcolors.com/hex-to-hsl to assist us in this task. Even more fortunately, the author has already completed this conversion. Here are the color codes represented in hsl() values:

```
--primary-color: #0c457d;
HSL: hsl(211, 84%, 26%)
--secondary-color: #e8702a;
HSL: hsl(20, 81%, 51%)
--text-color: #061b1a;
HSL: hsl(178, 100%, 6%)
--background-color: #fcf6ed;
HSL: hsl(45, 83%, 94%)
--highlight-color: #6bd2db;
HSL: hsl(183, 66%, 67%)
```

Let's start this exercise by following these steps:

1. Open the `light.css` file and change the colors in hexadecimal values to the preceding representation in `hsl()`:

```
:root {
    --primary-color: hsl(211, 84%, 26%);
    --secondary-color: hsl(20, 81%, 51%);
    --text-color: hsl(178, 100%, 6%);
    --background-color: hsl(45, 83%, 94%);
    --highlight-color: hsl(183, 66%, 67%);
}
```

2. Now, let's create another CSS file to set the colors for a dark theme. Create a new file in the root directory of the project (the same location as `light.css`) and name it `dark.css`. The structure should be like *Figure 6.8*:

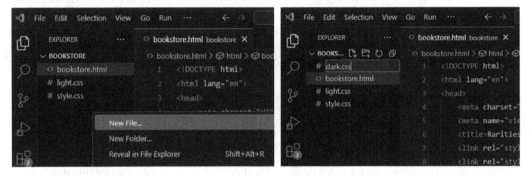

Figure 6.8 – Folder structure opened in VS Code

3. Well done! Now, we need to find the opposite colors to create the dark theme. To find the opposite colors (complementary colors) for the given colors in HSL format, we can add 180 degrees to the hue value. Here are the opposite colors for the provided colors:

```
--primary-color: hsl(31, 84%, 26%);
--secondary-color: hsl(200, 81%, 51%);
--text-color: hsl(358, 100%, 6%);
--background-color: hsl(225, 83%, 94%);
--highlight-color: hsl(3, 66%, 67%);
```

> **Note**
>
> These opposite colors are obtained by adding 180 degrees to the hue value of each original color.
>
> When finding the opposite color (complementary color) by adding 180 degrees to the original hue, the resulting value may exceed 360 degrees.
>
> However, in the HSL color model, values greater than 360 degrees "wrap around" the color wheel. This means that if the resulting hue value exceeds 360 degrees, it loops back to the beginning of the color wheel.

4. Create the variables as follows in the `dark.css` file:

```
:root {
  --primary-color: hsl(31, 84%, 26%);
  --secondary-color: hsl(200, 81%, 51%);
  --text-color: hsl(358, 100%, 6%);
  --background-color: hsl(225, 83%, 94%);
```

```
    --highlight-color: hsl(3, 66%, 67%);
}
```

5. In the HTML <head> section, establish a connection with your new CSS file by adding a
    new <link> tag to reference the dark.css stylesheet. Ensure that this <link> element is
    positioned after the one referencing light.css:

```
<head>
    <meta charset="UTF-8">
    <meta
        name="viewport"
        content="width=device-width,
                initial-scale=1.0"
    >
    <title>Rarities - Bookstore</title>
    <link rel="stylesheet" href="style.css">
    <link rel="stylesheet" href="light.css">
    <link rel=»stylesheet» href=»dark.css»>
</head>
```

Thanks to the cascade rule of CSS and the predefined color variables, the application has
seamlessly adopted the newly created theme, as shown in *Figure 6.9*.

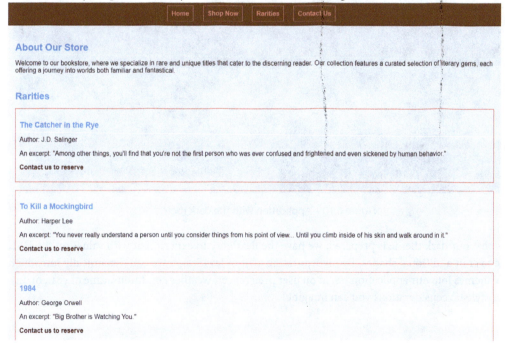

Figure 6.9 – Application with the new theme

As it turns out, our dark theme isn't as dark as we intended. This is because we haven't adjusted the lightness values in the `hsl()` function. Let's increase the lightness of the text and decrease it for the background to achieve a truly dark theme.

6.   Now, recalling that 0% lightness corresponds to black and 100% lightness to white in the `hsl()` logic, let's assign these values to the background and text variables in our `dark.css` file, respectively:

```
:root {
    --primary-color: hsl(31, 84%, 26%);
    --secondary-color: hsl(200, 81%, 51%);
    --text-color: hsl(358, 100%, 100%);
    --background-color: hsl(225, 83%, 0%);
    --highlight-color: hsl(3, 66%, 67%);
}
```

The result is *Figure 6.10*:

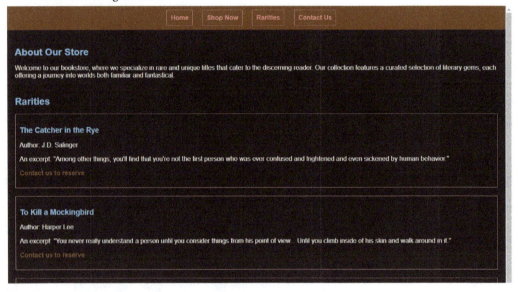

Figure 6.10 – Application with the dark theme

Now that our dark theme is prepared, we have the flexibility to experiment with values on the HSL scale to craft additional shades. By incorporating a bit of JavaScript, we can dynamically introduce these themes into our application based on user preferences, weather conditions, time of year, or any other stylistic considerations you can imagine!

Understanding the invert() filter

In the realm of dark theme design, CSS offers a powerful ally: the invert () filter. The invert () filter in CSS is used to invert the colors of an element, effectively producing a negative effect. It adjusts the hue of each pixel to its opposite on the color wheel, resulting in a color inversion.

The invert () filter function accepts a parameter, which represents the amount of inversion to apply. This parameter is a percentage, where 0% represents no inversion (i.e., the original colors), and 100% represents full inversion (i.e., completely inverted colors).

Here's an example of how you would use the invert () filter in CSS:

```
.element { filter: invert(100%); }
```

This applies 100% inversion (completely inverted colors).

You can also apply partial inversion by specifying a value between 0% and 100%, depending on the desired effect:

```
element { filter: invert(75%); }
```

This applies 75% inversion (partially inverted colors).

> **Important note**
> When it comes to the invert () filter, 50% means total neutrality, canceling any color, shadow, or brightness the element may have, as shown in *Figure 6.11*.

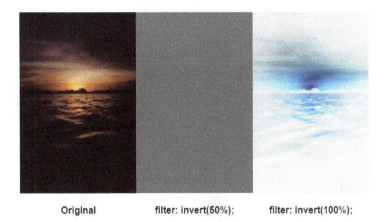

Original   filter: invert(50%);   filter: invert(100%);

Figure 6.11 – The invert() filter effect

The invert () filter can be a useful tool for creating visual effects, such as night mode themes or artistic alterations to images and elements on a web page.

Exercise 3 – creating a dark theme with the invert() filter

The `invert()` filter provides a simpler method for creating inverse themes, and we'll explore its application by refactoring our `dark.css` file to utilize this feature:

1. Please navigate to your project folder in VS Code and open the `dark.css` file, which may resemble the following:

    ```css
    :root {
      --primary-color: hsl(31, 84%, 26%);
      --secondary-color: hsl(200, 81%, 51%);
      --text-color: hsl(358, 100%, 100%);
      --background-color: hsl(225, 83%, 0%);
      --highlight-color: hsl(3, 66%, 67%);
    }
    ```

2. We'll begin by refactoring this code by removing all properties inside the `:root` selector:

    ```css
    :root {
    }
    ```

3. Next, we'll apply the `invert()` filter to the `:root` element to invert the colors of every element on the page:

    ```css
    :root {
      filter: invert(100%);
    }
    ```

4. The visual result should be looking like *Figure 6.12*:

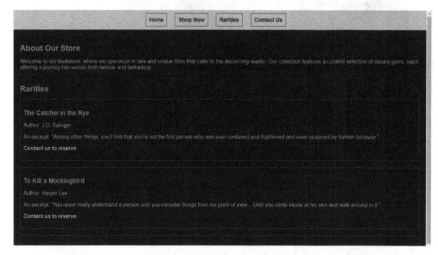

Figure 6.12 – Using the invert() filter to create a dark theme

> **There are two important considerations for this exercise**
>
> First, since the `invert()` filter affects not only hue but also saturation and lightness, there's no need to adjust these values to achieve a complete inverse theme.
>
> Second, if your application includes any images, their colors will also be affected. Therefore, it's essential to be specific about which elements you want to apply the `invert()` filter to in order to avoid unexpected results.

As we've discussed, the `invert()` filter can simplify theme styling. However, it's important to specify which elements should be affected by this function. Now, let's explore how the `color-scheme` property can enhance UX.

Exploring user's preferences with color-scheme

Since 2022, CSS has introduced a novel approach to dynamically setting color themes without the necessity of JavaScript, reflecting the web's growing trend toward customizable application themes.

The `color-scheme` CSS property enables an element to specify the color schemes in which it can be optimally displayed. The prevalent choices for operating system color schemes include "light" and "dark," also referred to as "day mode" and "night mode." Upon a user's selection of a preferred color scheme, the operating system implements modifications across the UI components, such as form controls and scrollbars, as well as adjusts the CSS system colors in use.

The `color-scheme` CSS property significantly enhances accessibility by ensuring that web content adapts seamlessly to the user's preferred color settings, reducing eye strain and making content more readable. For individuals with visual impairments or light sensitivity, the ability to automatically switch between light and dark modes can provide a more comfortable viewing experience. This adaptability supports a broader range of user needs without requiring manual adjustments, promoting an inclusive web environment. By aligning web applications with the user's system settings, developers can ensure that their content is not only aesthetically pleasing but also accessible to a diverse audience, contributing to a more equitable digital experience.

Here is the syntax:

```
element {
  color-scheme: [value]
}
```

The `color-scheme` property accepts the following values:

- `normal`: This signifies that the element does not recognize any color schemes and should display in the browser's standard colors.

- `light`: This denotes that the element is compatible with the operating system's light mode for visual presentation.

- `dark`: This implies that the element supports rendering in the operating system's dark mode color scheme.

- `only`: This prohibits the user agent from altering the element's color scheme. This can be utilized to deactivate color modifications triggered by Chrome's Auto Dark Theme, by setting `color-scheme: only light`, for a particular element or the `:root` selector.

How to declare color-scheme in CSS

To set the entire page with the user's color scheme preferences, specify the `color-scheme` property on the `:root` element:

```css
:root {
   color-scheme: light dark;
}
```

The preceding code facilitates dynamic CSS rendering depending on the browser's color scheme.

To align specific elements with the user's color scheme preferences, specify the `color-scheme` property on those elements:

```css
header {
   color-scheme: only light;
}
main {
   color-scheme: light dark;
}
footer {
   color-scheme: only dark;
}
```

How to style based on color schemes

To customize elements according to color scheme preferences, we utilize the `prefers-color-scheme` media query.

The following example demonstrates opting the entire page into both light and dark operating system color schemes using the `color-scheme` property.

Additionally, it employs `prefers-color-scheme` to define the preferred foreground and background colors for individual elements within those schemes:

```css
:root {
   color-scheme: light dark;
}
```

```css
@media (prefers-color-scheme: light) {
  body {
    color: black;
    background-color: white;
  }
}

@media (prefers-color-scheme: dark) {
  body {
    color: white;
    background-color: black;
  }
}
```

CSS color schemes have introduced a new way for developers to leverage user preferences, ensuring a seamless and attractive experience. Let's explore another option for styling based on user preferences: the light-dark() function.

Introducing the light-dark() function

A groundbreaking addition to CSS, the light-dark() function simplifies the process of establishing dynamic themes. This function enables the setting of foreground and background colors for various color schemes in just one line of code, eliminating the need for the prefers-color-scheme media queries.

Here is the syntax for this:

```css
element {
  property: light-dark(
    [color for light or default],
    [color for dark]
  );
}
```

The light-dark() CSS color function returns the first value if the user's preference is set to light or if no preference is set, and the second value if the user's preference is set to dark:

```css
:root {
  color-scheme: light dark;
}

.element {
  color: light-dark(black, white);
  background-color: light-dark(white, black);
}
```

> **Important: `light-dark()` is an experimental feature**
>
> Please note that this feature is experimental and may have limited availability. It's recommended to test it in your browser environment.

Exercise 4 – simplifying our CSS theming with media queries

To celebrate our journey into the world of CSS theming, let's practice once more with our bookstore application. We're going to refactor the CSS theming to set it dynamically based on the user's preferences. For testing purposes, we'll use the Microsoft Windows operating system as an example:

1. Start by opening the project folder in VS Code and creating a new file named `color-variation.css` in the same location as the other CSS files. We'll use a single CSS file for both light and dark themes to reduce code repetition.

2. After creating the file, let's define that our `:root` element (representing the whole HTML) will accept variations of light and dark modes. We'll set this writing the following notation in `color-variation.css`:

    ```css
    :root {
      color-scheme: light dark;
    }
    ```

3. After setting this up, let's create a media query to configure the colors for light mode. We're creating the `prefers-color-scheme` media query in `color-"variation.css"` and pasting the entire content from `light.css` inside it:

    ```css
    @media (prefers-color-scheme: light) {
      :root {
        --primary-color: #0c457d;
        --secondary-color: #e8702a;
        --text-color: #061b1a;
        --background-color: #fcf6ed;
        --highlight-color: #6bd2db;
      }
    }
    ```

4. Great! We're halfway there. Now, let's do the same with the `dark.css` file. Create a `prefers-color-scheme` media query and paste all the content from `dark.css` inside it. Note that we're not using the `invert()` filter because, in this context, when one theme is selected, the browser does not render the other one, so `invert()` does not have anything to invert from:

    ```css
    @media (prefers-color-scheme: dark) {
      :root {
        --primary-color: hsl(31, 84%, 26%);
        --secondary-color: hsl(200, 81%, 51%);
    ```

```
    --text-color: hsl(358, 100%, 100%);
    --background-color: hsl(225, 83%, 0%);
    --highlight-color: hsl(3, 66%, 67%);
  }
}
```

The resulting CSS file should look like this:

```
:root {
  color-scheme: light dark;
}

@media (prefers-color-scheme: light) {
  :root {
    --primary-color:    #0c457d;
    --secondary-color: #e8702a;
    --text-color: #061b1a;
    --background-color: #fcf6ed;
    --highlight-color: #6bd2db;
  }
}

@media (prefers-color-scheme: dark) {
  :root {
    --primary-color: hsl(31, 84%, 26%);
    --secondary-color: hsl(200, 81%, 51%);
    --text-color: hsl(358, 100%, 100%);
    --background-color: hsl(225, 83%, 0%);
    --highlight-color: hsl(3, 66%, 67%);
  }
}
```

5. Now, it's time to connect our new CSS file to our HTML. Delete the `<link>` elements that refer to `light.css` and `dark.css` and replace them with the new file that contains all the coloring and logic for our dynamic theme, `color-variation.css`:

```
<head>
    <meta charset="UTF-8">
    <meta
        name="viewport"
        content="width=device-width,
                initial-scale=1.0"
    >
    <title>Rarities - Bookstore</title>
    <link rel="stylesheet" href="style.css">
```

```
        <link rel=»stylesheet» href=»color-variation.css»>
    </head>
```

6. After that, let's configure our operating system to prefer the dark theme, so we can test it properly.

    On Windows 11, right-click with your mouse on the desktop and select **Personalization** | **Colors**. Open **Colors setting** and, under **Choose your color**, select **Custom**. Under **Choose your default app mode**, select **Dark**.

    All set! Open the application with your browser to see the result.

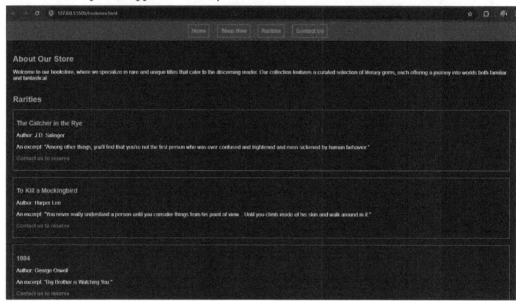

Figure 6.13 – How the application should look in the browser

That's it! We've created an application that changes themes *dynamically* with no need for JavaScript, just with CSS. Theming extends beyond mere color choices; typography is equally vital in shaping a well-designed experience. Neglecting typography can detract from an otherwise thoughtfully curated theme, highlighting the significance of this aspect, which we'll explore in the next section.

Understanding typography and font choices

Typography plays a crucial role in web design as it directly affects readability, UX, and the overall aesthetics of a website. When choosing the best font for your theme, consider the mood and message you want to convey, the styling and brand definitions, as well as the preferences and expectations of your target audience. Here's a guide to choosing good fonts for the application:

- **Legibility and readability**: Legibility refers to how easily one can distinguish individual characters, while readability refers to how easy it is to read blocks of text. Choose fonts that are clear and easy to read, especially for longer passages of text.

  Avoid overly decorative or ornate fonts for body text, as they can be difficult to read in smaller sizes. Ensure sufficient contrast between text and background colors to improve readability, especially for users with visual impairments. In the following figure, observe how varying font families, sizes, and contrast levels impact legibility.

Figure 6.14 – Depiction of various legibility scenarios in high, medium,
and low contrast between font and background color

- **Font families**: Fonts are generally categorized into serif, sans-serif, monospaced, and display fonts.

  Serif fonts have small decorative lines at the end of strokes, while **sans-serif fonts** do not. **Monospaced fonts** have equal spacing between characters, while **display fonts** are more decorative and suitable for headings or titles.

  Sans-serif fonts are often preferred for body text on the web due to their clean and modern appearance, but serif fonts can work well for certain themes or when aiming for a more traditional look. In the following figure, we'll explore an example of each font family element:

  Times New Roman is a serif font

  Arial is a sans-serif font

  Courier New is a monospaced font

  Vivaldi is a display font

Figure 6.15 – Representation of different font families

- **Consistency**: Maintain consistency in typography throughout your website to create a cohesive look and feel. Choose a primary font for headings and another for body text, and stick to these choices across the site. Use font weights and styles (such as bold, italic, or regular) consistently to provide visual hierarchy and emphasis.

In *Figure 6.16*, take note of how font consistency plays a pivotal role in crafting harmony throughout the website. Pay close attention to the chosen font families, font colors, and font weights, as they collectively narrate a story and elevate the overall UX.

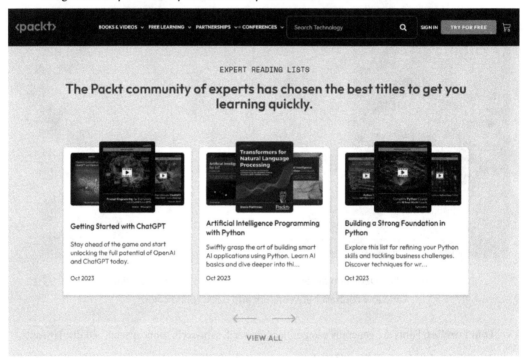

Figure 6.16 – The Packt home page

- **Hierarchy**: Create a visual hierarchy by using different font sizes, weights, and styles for headings, subheadings, and body text. Larger and bolder fonts typically indicate more important information. *Figure 6.17* is an example illustrating how font hierarchy operates:

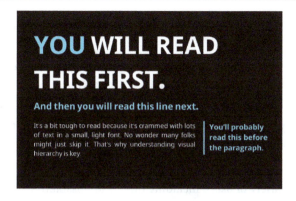

Figure 6.17 – Font hierarchy representation

- **Brand personality**: choose fonts that reflect the personality and tone of your brand or theme. A sleek and modern font might be suitable for a tech-related website, while a vintage-inspired font could work well for a retro-themed site. Fonts have the ability to evoke emotions and this must be considered during development, as can be seen in *Figure 6.18*:

Figure 6.18 – Illustration of how different font choices impact the message's meaning

It's incredible how CSS has evolved! Modern CSS enables us to craft stunning themes that deliver the precise experience we've envisioned for our users. It also provides us with more tools to create accessible applications, as we will explore next.

Accessibility and inclusive design considerations

When considering accessibility and inclusive design in CSS theming and font selection, there are several key considerations to keep in mind:

- **Contrast ratio**: Ensure sufficient contrast between text and background colors to make content readable for users with visual impairments. Use tools such as the WCAG contrast checker to verify that your chosen color combinations meet accessibility standards.

- **Font size and scalability**: Choose fonts and set font sizes that are legible and resizable without loss of clarity. Use relative units such as percentages or rem to allow users to adjust text size according to their preferences.

- **Font weight and style**: Avoid using overly thin or light fonts, as they may be difficult to read for some users.

- **Font family**: Select font families that are versatile and support a wide range of characters, including those from different languages and writing systems.

- **Screen reader compatibility**: Use semantic HTML elements and ARIA attributes to enhance the accessibility of your content for screen reader users. Provide descriptive alt text for images and use proper heading structures to facilitate navigation.

Consider the diverse needs and preferences of users when selecting fonts and designing CSS themes. Test your designs with a variety of users, including those with disabilities, to identify and address potential barriers to accessibility.

Modern CSS enables us to create more interactive and fluid applications with lower performance costs than JavaScript, which is a great strategy. As you continue exploring CSS, remember the importance of the font selection strategies and polishing techniques we've covered.

I encourage you to delve deeper into modern CSS concepts of themes, variables, and dynamics. *Learn by doing* and build an amazing portfolio!

Summary

In this chapter, we delved into the creation of dynamic themes using CSS color variables, the `invert()` filter, and the innovative `color-scheme` property. Understanding the mechanics of HSL colors was essential, along with adopting best practices for building CSS code that remains maintainable and non-redundant. We explored the exciting possibilities of modern CSS through `prefers-color-scheme` media queries and the `light-dark()` CSS function.

We also explored font selection, considering factors such as legibility, UX, hierarchy, and brand styling, while also learning various polishing techniques to refine our designs and enhance the overall UX.

As we fortified our knowledge in web development, we honed our skills to craft more powerful, performant, and user-friendly applications. With this solid foundation in CSS theming and color variables, I anticipate that you'll embark on creating even greater and more interactive projects.

Furthermore, our exploration of CSS theming not only enhances the aesthetics and UX but also sets the stage for our next chapter, *Using CSS and HTML to Boost Performance*. Building upon our understanding of CSS variables, media queries, and best practices, we'll now shift our focus to maximizing the performance of our web applications through strategic CSS and HTML optimization techniques.

Part 3:
Building for All

Creating quality code goes beyond crafting visually appealing websites; it involves ensuring that everyone can access and enjoy a consistent experience with your application. In this section, we'll learn how to build websites that are accessible and responsive, reaching a wider audience regardless of their circumstances. We'll start by enhancing HTML and CSS to boost performance, followed by techniques to achieve responsiveness. We'll also cover accessibility in depth. These three pillars form the foundation of a robust web application that delivers a superior user experience, regardless of the user's device or condition.

This part contains the following chapters:

- *Chapter 7, Using CSS and HTML to Boost Performance*
- *Chapter 8, Responsive Web Design and Media Queries*
- *Chapter 9, Ensuring Accessibility in HTML and CSS*

7

Using CSS and HTML to Boost Performance

Often, we digital creators work on the latest laptops and test on the latest high-end devices, but the audience of a website can be broad and very diverse. Downloading thousands of high-quality images for a web photo gallery might be quick over super-fast fiber optic broadband, but on a mobile network, it might use a great chunk of the user's data contract and be a painful experience.

Considering the performance of a website can be the difference between success and failure. As websites have become more sophisticated, with richer interactions and more complexity, they have also become heavier, with more and more data and assets to load.

In this chapter, we're going to cover the following main topics:

- Understanding the performance of a web page
- Measuring performance with Lighthouse
- Improving the performance of a web page

First, we will look at what we mean by performance in a web context and how we can go about measuring performance, so we know what changes make effective improvements to the quality of the experience for all our users.

Technical requirements

The code files for this chapter can be found at https://packt.link/X9zRL.

Understanding the performance of a web page

Web Vitals is an initiative promoted by Google to standardize the measurement of the user experience and performance of a website. One aim of the initiative is to determine the most important of these performance measures, which they call the *Core Web Vitals*.

The metrics defined and used to measure web performance evolve over time but, since 2020, the three main metrics used are as follows:

- **Largest Contentful Paint (LCP)**
- **First Input Delay (FID)**
- **Cumulative Layout Shift (CLS)**

Each metric measures a different aspect of user experience as it is impacted by performance.

By following best practices and getting these aspects of a web page right, you will not only improve the performance a user will experience but also the chance of your web page being rated highly by Google, and therefore improving SEO.

Let's look at the metrics in detail.

Largest Contentful Paint (LCP)

LCP measures the loading performance of a web page by evaluating the time taken for the largest element on the page to load. The element in question could be an image, a heading, or a block of text, and the measurement is complete when that element becomes visible within the user's viewport.

The goal of LCP is to keep the time it takes for the largest element to be visible to a minimum. The recommendation for a good experience is anything under 2.5 seconds; anything over 4 seconds is considered poor; and anything between those two values needs improvement.

An example might be a hero image on a news site that takes up a large portion of the viewport above the fold. By measuring how long that image takes to load, we can get an insight into how quickly users can see the main content of the page. If that image takes 1.5 seconds to load, then the LCP of the page would be considered good.

LCP is an important part of user experience because it corresponds to the perceived speed of a web page. The faster the LCP, the faster users can quickly access and engage with the primary content of a page. This leads to a satisfying user experience. A slow LCP can result in users feeling the page is slow to load and frustrating to use.

To improve the LCP of a web page, we can optimize assets and make sure the critical resources on our web page are prioritized to load the quickest.

Cumulative Layout Shift (CLS)

CLS measures the visible stability of a web page as it loads. It evaluates the amount of unexpected layout shifts that occur during page rendering.

CLS is given as a score calculated by multiplying the impact fraction (the amount of the viewport that is affected by the shift) by the distance fraction (the distance the elements shift relative to the viewport).

The goal of CLS is to keep shifts in layout to a minimum. The recommendation for a good experience is a score below 0.1; anything above 0.25 is considered poor; and anything in between those two values needs improvement.

Layout shifts are often disruptive to user experience. An example might be on a news website when an advert pushes an article down the page, causing the user to lose their position or a form where you might unintentionally interact with the wrong element when a button shifts position.

It is important to note that a page can still change dynamically and score well for CLS. Layout shift is calculated based on unintentional shifts, that is, shifts not caused by user interactions.

We can improve the CLS of a web page by making sure images and embedded content (such as ads or an iframe) have dimensions set so their position and size on the page can be calculated during loading. We can also use placeholders or loading skeletons for late-loading elements. In some cases, gradual transitions and animations can help.

First Input Delay (FID)

FID is a measurement of the interactivity and responsiveness of a web page. It does this by measuring the time it takes for a web page to respond to the first user interaction, such as clicking a button or selecting a menu item.

The goal is to reduce the delay between user input and the browser's response to provide a seamless and enjoyable user experience.

The goal of FID is to keep interaction responsive times fast. The recommendation for a good experience is a response speed of under 100 ms; anything above 300 ms is considered poor; and anything in between those values needs improvement.

A poor FID score can create a very frustrating experience for a user, with a sense that the page is sluggish to respond. It can cause frustration and high drop-off rates among users.

Heavy use of JavaScript in a web page is often the cause of slow FID. We can improve the FID of a page by minimizing long tasks on the main thread, optimizing and prioritizing critical JavaScript, and using web workers to offload complex tasks to separate threads, ensuring a more responsive user interface.

> **Note**
>
> FID was one of the Core Web Vitals up until March 2024 when a new metric called **Interaction to Next Paint** (**INP**) replaced it. INP is considered a better measurement of the responsiveness and interactivity of a web page.

Interaction to Next Paint (INP)

INP is a new measurement that replaced FID as part of the Core Web Vitals in March 2024.

Like FID, it is a measurement of how quickly a page responds to user interactions. However, the measurement is different – it observes the latency of click, tap, and keyboard interactions over the lifespan of a page. The concern is less with the first impression of a first interaction and more with the whole experience of interacting with the page.

Now that we have looked at the core web metrics for understanding web performance, in the next section, we will use the Lighthouse auditing tool to generate a report. Many of the results in the Lighthouse performance report are based on the metrics we have described.

Measuring performance with Lighthouse

Lighthouse is an auditing tool that can generate reports for several categories of optimizations that we might want to make to web page behavior. These categories are Performance, Best Practices, Accessibility, SEO, and Progressive Web Apps.

While Lighthouse is not exclusively a web performance tool, it is useful for giving us a headline score for how our web page loads and performs, and a simple report of issues that users may be facing with our web page.

Lighthouse is a widely used tool that will audit a web page and return a report with a score for performance and a list of issues and pointers for improving that score.

The great thing about Lighthouse is that it is easy to use. It is available through lots of tools; for example, you can use it from the command line, or you can use it to generate reports in a Node.js application or as part of a continuous integration pipeline. By far the easiest way for us to access it as web developers is through the web browser.

Lighthouse is available via Chrome's web developer tools, so we don't have to download any additional software, and that means we can access it at any time while working on a web page.

The following steps will help you to do a performance audit with Lighthouse in the Chrome browser:

1. Open the web page you want to run an audit on.
2. Open the Chrome web developer tools.
3. Select the **Lighthouse** tab.

4. From the options on the **Lighthouse** panel, deselect all the categories except **Performance**.

5. Click the **Analyze page load** button, as shown in *Figure 7.1*.

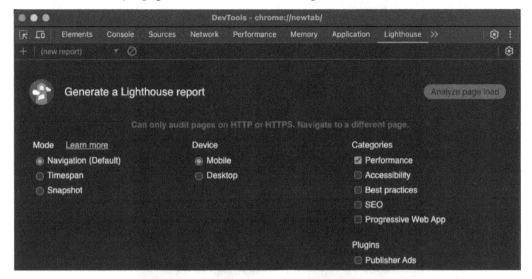

Figure 7.1: Lighthouse tab in Chrome developer tools

Lighthouse will take a bit of time to run an audit on the current web page. This can take about 30-60 seconds, depending on the size of the page and the options selected. Once the audit is complete, you will see a report for the page.

> **Tip**
>
> For the best results, it is recommended to run Lighthouse in an incognito browser window. This makes the results more predictable because no extensions will be running, which could affect the reported score.

Anatomy of the Lighthouse report

A Lighthouse performance audit report has three main sections: **Score**, **Metrics**, and **Diagnostics**.

Score

This is a percentage score calculated to give an overall sense of how your web page performs for a given category (e.g. Performance, SEO, and Accessibility). In the case of Performance, the score is a weighted average based on several web performance metrics, including some of the ones we have already described in this chapter.

As shown in *Figure 7.2*, the scores are color-coded using a traffic light system, with a red band for a poor score (indicating that the performance will deeply affect a user's experience), orange representing a need for improvement, and green being a good score that will leave our users with a good experience.

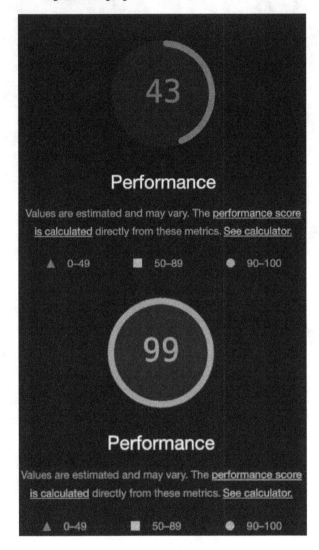

Figure 7.2: Examples of a bad score (43%) and a good score (99%) in Lighthouse

Metrics

The next section of the report is a breakdown of the score into its component metrics.

For performance, the five metrics used are the Core Web Vitals related to page load. We described CLS and LCP earlier in the chapter. The other three are **First Contentful Paint (FCP)**, **Total Blocking Time (TBT)**, and Speed Index.

FCP is a measure of perceived load speed. By measuring the time taken before the first element (text or image) is painted on the screen, we get an idea of whether the user will feel the page is loading or has perhaps stalled.

TBT measures the time from FCP until the page can be interacted with. If a page has rendered some content (such as a loader) but interactivity is still blocked for a very long time, it can be harmful to the user's experience.

Speed Index measures how long it takes for the content of a web page to load and compares it to data on other real websites.

Each metric will have a value categorized as green (good), orange (needs improvement), or red (poor):

- A good CLS score is under 0.1, with a value between 0.1 and 0.25 needing improvement, and anything above 0.25 being considered poor. See the earlier section on CLS for more information on how this score is calculated.

- A good LCP score is under 2.5 seconds, with needs improvement being a value between 2.5 seconds and 4.0 seconds, and anything greater than 4.0 seconds being considered poor.

- FCP is another indicator of how a user perceives load speed. It measures the time before the first text or image is painted. The value is given in seconds, with a score over 3.0 seconds being poor, a score between 3.0 seconds and 1.8 seconds needing improvement, and a score under 1.8 seconds being good.

- TBT is another page load metric. It is based on a measure of the amount of time after FCP where input responsiveness is blocked. A time of 0-200 ms is considered good. 200-600 ms will need improvement, and anything above that is considered poor.

- Speed Index measures how quickly the contents of a page are visually displayed as the page loads. It is a time measurement with anything less than 3.4 seconds considered good. 3.4 seconds to 5.8 seconds needs improvement, and anything over 5.8 seconds is considered poor.

In the following screenshot, we can see the metrics for a page with a mixture of poor (FCP and LCP), needs improvement (TBT and Speed Index), and good performance (CLS) scores.

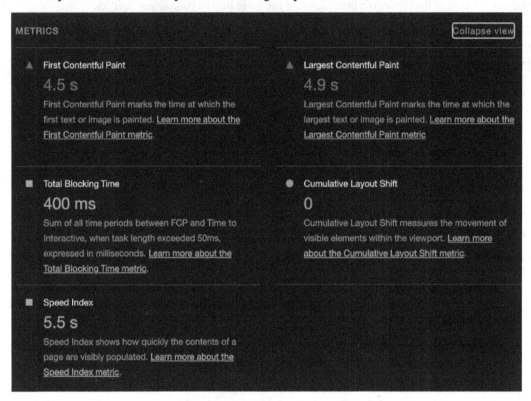

METRICS   Collapse view

▲ First Contentful Paint   ▲ Largest Contentful Paint

4.5 s   4.9 s

First Contentful Paint marks the time at which the   Largest Contentful Paint marks the time at which the
first text or image is painted. Learn more about the   largest text or image is painted. Learn more about the
First Contentful Paint metric.   Largest Contentful Paint metric

■ Total Blocking Time   ● Cumulative Layout Shift

400 ms   0

Sum of all time periods between FCP and Time to   Cumulative Layout Shift measures the movement of
Interactive, when task length exceeded 50ms,   visible elements within the viewport. Learn more
expressed in milliseconds. Learn more about the   about the Cumulative Layout Shift metric.
Total Blocking Time metric.

■ Speed Index

5.5 s

Speed Index shows how quickly the contents of a
page are visibly populated. Learn more about the
Speed Index metric.

Figure 7.3: An example of the METRICS section of a Lighthouse report

Diagnostics

The **DIAGNOSTICS** section of the report drills further into the details of any problems and flags opportunities where we could improve performance. This includes, for example, listing elements that are causing layout shifts or images that are unoptimized and take a long time to load.

In the following screenshot, we can see the diagnostics results with suggestions for areas to improve and potential problem points that affect the performance of the web page. In this example, elements that cause layout shift are flagged, as well as unoptimized images and an element with a very poor LCP.

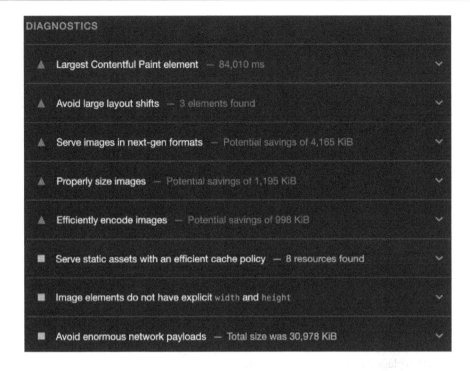

Figure 7.4: An example of the DIAGNOSTICS section of a Lighthouse report

In the following exercise, we will set up a web page that has some performance issues, and we will run a Lighthouse performance audit on it so we can examine the results and get an understanding of how decisions we make while developing a web page can impact its performance score.

Exercise – finding performance issues in a web page

In this exercise, we will demonstrate how to use a Lighthouse audit to better understand optimizations we can make to our web page's performance.

We will create a web page with a simple image gallery using HTML and CSS. This gallery will not be optimized for performance and will, in fact, have some fundamental performance issues that we will fix as we progress through the rest of the chapter.

Here are the steps to create a web page with an image:

1. Create a new web page to run a performance audit on in Visual Studio Code (**File > New File...**). Let's name the file exercise_01.html and save it in a folder called chapter_7.

2. To have something to audit with Lighthouse, we will create a very simple photo gallery, which will be the basis for this exercise and the rest of the exercises in this chapter. The gallery will use CSS and HTML, with no JavaScript involved. To create the gallery, we will start with a simple HTML document:

```
<!DOCTYPE html>
<html lang="en">
  <head>
    <meta
      name="viewport"
      content="width=device-width, initial-scale=1">
    <title>A gallery</title>
  </head>
  <body>
    <div class="gallery">
      <!-- insert gallery html here -->
    </div>
  </body>
</html>
```

3. The gallery will have two main parts – a large hero image and a list of thumbnails that can be selected to replace the hero image. As we aren't using any JavaScript, we will need to represent all the images in the gallery declaratively in the HTML document. For the hero images, we add the following code to the `div` element with the `.gallery` class:

```
<div class="full">
  <img
    id="hero-1"
    src="assets/img-01.jpg"
    alt=""
    class="hero image"
  />
  <img
    id="hero-2"
    src="assets/img-02.jpg"
    alt=""
    class="hero image"
  />
  <img
    id="hero-3"
    src="assets/img-03.jpg"
    alt=""
    class="hero image"
  />
```

```
    <img
      id="hero-4"
      src="assets/img-04.jpg"
      alt=""
      class="hero image"
    />
    <img
      id="hero-5"
      src="assets/img-05.jpg"
      alt=""
      class="hero image"
    />
    <img
      id="hero-6"
      src="assets/img-06.jpg"
      alt=""
      class="hero image"
    />
    <img
      id="hero-7"
      src="assets/img-07.jpg"
      alt=""
      class="hero image"
    />
    <img
      id="hero-8"
      src="assets/img-08.jpg"
      alt=""
      class="hero image"
    />
    <img
      id="hero-9"
      src="assets/img-09.jpg"
      alt=""
      class="hero image"
    />
    <img
      id="hero-10"
      src="assets/img-10.jpg"
      alt=""
      class="hero image"
    />
</div>
```

4. The thumbnails will similarly be represented by a list of each of the images in the gallery. Below the hero images, we add an unordered list to create the thumbnail list. For each thumbnail, we will add a list item with an anchor and an image. The entire code can be found in the GitHub repo (https://packt.link/X9zRL):

```
<ul class="thumbnails">
  <li>
    <a href="#hero-1">
      <img
        src="assets/thumb-01.jpg"
        alt=""
        class="thumbnail image"
      />
    </a>
  </li>
</ul>
```

5. The hero image will take up most of the viewport. We'll use CSS to create the gallery layout and to style the thumbnail and hero images. We will add the style to the head of the HTML document. The entire code can be found in the GitHub repo.

6. To switch between each thumbnail image without navigating between individual web pages or using JavaScript, we'll use a little-known CSS pseudo-class called :target. It is used to select an element based on an ID matching the URL's hash. We will use the powerful new :has selector with :not to determine whether there is no element selected by :target and, in that case, we will show the first hero image. We won't go into details about using these selectors here, but for more information, see the MDN entries: https://developer.mozilla.org/en-US/docs/Web/CSS/:target, https://developer.mozilla.org/en-US/docs/Web/CSS/:has, and https://developer.mozilla.org/en-US/docs/Web/CSS/:not.

```
.full:not(:has(.image:target))
  .image:first-of-type,
.full .image:target {
  z-index: 1;
  display: block;
}
```

7. Finally, we'll add a loader to the gallery that is shown while the hero image is loading. We add a div element to the hero images, <div class="loading"></div>, and the following additional styles:

```
.loading {
    width: 3rem;
```

```
        height: 3rem;
        border: .375rem solid lightskyblue;
        border-color:
            transparent lightblue lightblue
            lightblue;
        border-radius: 50%;
        position: absolute;
        animation:
            rotating 1s
            cubic-bezier(0.33, 0, 0.33, 0.9)
            infinite;
    }

    @keyframes rotating {
        from {
            rotate: 0;
        }
        to {
            rotate: 360deg;
        }
    }
```

In the following screenshot, we can see what the gallery we have created should look like, with a large hero image taking up most of the viewport and a row of thumbnails below:

Figure 7.5: The gallery we have created in this exercise

8.   Now that we have created our gallery, we will run a Lighthouse audit to get an idea of how the web page performs and to see if there are any areas we can improve. In Chrome, open the developer tools, select the **Lighthouse** tab, deselect all categories except **Performance**, and click the **Analyze page load** button.

The following screenshot shows the options selected prior to clicking the **Analyze page load** button to run the performance audit on the gallery.

Figure 7.6: The Lighthouse report settings page with the Performance category selected

The resulting report from Lighthouse should look similar to the following screenshot. We can see that the page gets a score of **46** for **Performance**, which is considered poor. Looking at the individual metrics, we can see that LCP, TBT, and CLS are particularly poor. Furthermore, the diagnostics suggestions point to issues with unoptimized images, layout shifts caused by the thumbnails and the loading indicator, and a detrimental LCP due to the hero image.

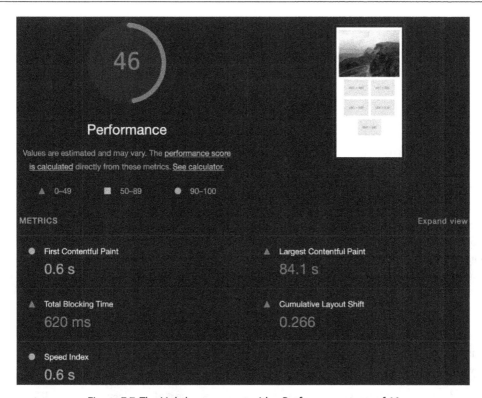

Figure 7.7: The Lighthouse report with a Performance score of 46

We'll come back to the gallery created in this exercise in the rest of this chapter as we solve some of the performance problems presented in this report. We'll look at a few solutions and apply them to the gallery to make a gradual improvement to the user experience of our gallery.

We've seen how we can measure the performance of a web page using Lighthouse. In the next section, we will make improvements to the gallery we've created to look at how they can impact performance and enhance the user experience.

Improving the performance of a web page

In this section, we will look at some of the mechanisms available in HTML that allow us to optimize and tweak how resources load on a web page. Controlling the load according to your specific requirements, making sure the right assets load upfront when needed, and loading less urgent resources with lower priority can help us get a rendered web page faster, improving our FCP and LCP scores.

Controlling how assets load

Images can bring a web page to life but, equally, they can bring a web page to a grinding halt while thousands of unoptimized images, amounting to many MB of data, are downloaded over a slow or intermittent connection.

> **Info**
>
> According to the Web Almanac state of the web report for 2022, the median page weight for a mobile website was 2,019 KB, with images contributing 881 of those KB (`https://almanac.httparchive.org/en/2022/page-weight#page-weight-by-the-numbers`).

When a browser parses an HTML document, it will do so starting from the top of the document and working down to the bottom of the document. When it comes across an `img` tag with a `src` attribute, the default behavior for the browser is to immediately create a connection to wherever that image asset is stored and to begin loading that image. When the image is loaded, it will be displayed.

Eager loading

This loading behavior is known as eager loading. In many cases, this is the behavior you want. If the image is the first thing the user will see, you want it to load as quickly as possible. However, there are times when eagerly loading all the images on a web page is not optimal.

Let's consider the example in the previous exercise, where we created a gallery of images. When the page loads, we see one main image and a row of thumbnails, and there are further thumbnail images that do not appear on screen until we have scrolled further down the page. Each of the thumbnail images can be clicked on to replace the main hero image with a larger version of the thumbnail image.

For a user arriving at the gallery, the most important thing will be to see the hero image, and we should load this as a priority. We don't want all the images to load with the same priority because some will be below the fold or will not be seen until they have been selected.

Lazy loading

One of the great benefits of modern web technologies is that they provide very simple declarative ways of controlling your web page's performance. A good example of that is the lazy loading attribute for images. By simply adding the `loading="lazy"` attribute, we have a mechanism to delay the loading of images while they are out of view.

If we want to stop an image loading until it is onscreen, we simply add the `loading="lazy"` attribute:

```
<img
  src="assets/important-image.jpg"
  loading="eager"
  alt=""
```

```
/>
<!-- somewhere lower down the page -->
<img
  src="assets/secondary-image.jpg"
  loading="lazy"
  alt=""
/>
```

Prioritizing loading

To optimize a web page's performance and the time taken to get to FCP, it is useful to optimize the critical path by loading critical assets as quickly as we can.

We've learned how to postpone the loading of an asset until it is needed by the user, but we also have some tools available to help us prioritize how assets are loaded. We can push them to the front of the queue with resource hints and fetch priority hints.

With resource hints, we set a `link` element with one of three `rel` attribute values – `preload`, `prefetch`, or `preconnect`. These values have the following effect on resource loading:

* `<link rel="preload"` is used to preload an external resource that is required for the initial render of the page

* `<link rel="prefetch"` is used to load content that will rendered on the next page

* `<link rel="preconnect"` establishes a connection to a server to prepare the connection for loading resources

We can specify the content type being preloaded with the `as` attribute. We can specify many content types. Some of the more commonly used are image, font, style, script, and fetch. So, for example, if we want to preload a particular image asset, we set the `as` attribute to `"image"`.

Priority hints give us further control over prioritizing resources by letting us set a hint for the browser to either increase or decrease the priority of a resource.

Priority hints are set using the `fetchpriority` attribute with one of three possible values: `auto`, `high`, or `low`. We can apply this attribute to resources via `img` tags or `link` tags that reference external resources, and we will prompt the browser to prioritize the resource differently.

Improving image performance

When it comes to images, fonts, and other heavy media assets, a lot of our performance issues can come down to the sheer size of the file. It is important for user experience to make sure we have optimized any images on our web page.

Things to consider when working with images on the web are whether we are using the best encoding and format for the image. All photos used to be `jpeg` and `png`, and these formats were used for lots of assets, but in recent years we have had a lot of improvements to web image formats – `webp`, `avif`, and `jpeg xl` all provide formats that can greatly reduce the size of an image.

> **Note**
>
> A great tool for trying out different compression techniques and settings for images is Squoosh (`https://squoosh.app/`).

Layout shift and images

Images can contribute to page load time. They can also contribute to another aspect of web performance – layout shift.

Images can cause layout shifts if the size of the element into which they are loaded is not the same as the size or aspect ratio of the image, and when the image loads it causes a shift in the layout of the page.

The important thing to consider when solving layout shifts with images is to set the dimensions for an image using width and height. This can be combined with the aspect ratio attribute.

Exercise – optimizing the performance of a web page

In this exercise, we will carry on from the previous exercise, where we created and tested the performance of an image gallery.

Here, we will do some work to optimize the performance of the image gallery we created using the techniques we have discussed for lazy loading images and preloading high-priority assets.

Here are the steps for optimizing the performance:

1. First, we will make a copy of the gallery. In the `chapter_7` folder that we created in the previous exercise, copy `exercise_01.html` and save it as `exercise_02.html`.

2. Our first task will be to lazily load all the images except for the first hero image. This will really help with the FCP metric. We will add the `loading="lazy"` attribute to each of the hero images (except for the first one) and to each of the thumbnail images. This way, the only image that will load if not onscreen is the first hero image.

3. Next, we'll hint to the browser which are our highest priority images, and which can be loaded later. Again, we will consider all the thumbnails and all hero images aside from the first hero image as a lower priority. We will set the `fetchpriority="low"` attribute for each of these images. We will then set the `fetchpriority="high"` attribute for the first hero image.

4. Finally, to make sure the first hero image loads prior to the first page render, we will add a preload link to the header of the page. Above the `style` tag in the head of the document, we add the following line of code:

```
<link
  rel="preload" as="image"
  href="assets/andrew-ridley-Kt5hRENuotI-unsplash.jpg"
/>
```

With these easy-to-implement optimizations, we will see some striking improvements to our web page loading experience. Testing in Lighthouse led to some real improvements in the overall performance score and the FCP and LCP scores.

While these scores can vary, you can see the results in the following screenshot with an overall score of 85% and FCP and LCP both showing **0.6 s**.

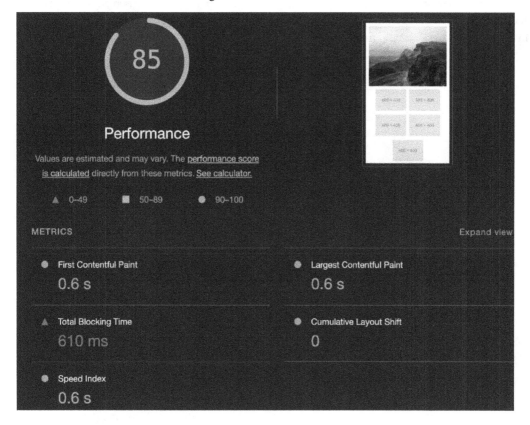

Figure 7.8: The Lighthouse report with a Performance score of 85

In this exercise, we've seen some simple ways to improve performance scores that, combined, can really improve the user experience of your web page.

Summary

In this chapter, we have learned how to use the popular **Lighthouse** tool to measure the performance of our web pages. We've learned about the Core Web Vitals – a set of important metrics for a user's experience of a web page. We've introduced techniques to lazily load some of the larger assets on a web page, and we've learned about prioritizing the order in which assets load.

The techniques we've looked at in this chapter are some simple browser features that we can use to improve the way our web pages load. This will benefit all users and improve engagement with your web pages.

In the next chapter, we'll look at how we can improve the user experience of our web pages for different devices and viewport sizes using responsive web design and media queries: techniques that can create an even better experience and further boost performance on mobile.

8

Responsive Web Design and Media Queries

When was the last time you accessed the internet? Was it via your mobile phone or a computer screen? In today's digital age, it's highly likely that you used a mobile device. As mobile internet usage continues to surpass desktop browsing, the importance of web responsivity becomes more critical than ever.

Historically, web design has evolved significantly with the proliferation of mobile devices. However, creating responsive web applications encompasses much more than just reducing and rearranging elements on a screen. It involves crafting an adaptable and seamless user experience across a diverse range of devices and screen sizes.

Understanding and implementing responsive design is not without its challenges. Developers must consider various screen sizes, resolutions, and orientations, all while maintaining performance and aesthetics. Despite these challenges, the benefits are clear: improved user satisfaction, higher engagement rates, and broader accessibility.

In this chapter, we're going to cover the following main topics:

- What does it mean to design mobile first?
- Viewports
- Media queries
- Responsive values in CSS
- Responsive images
- CSS Flexbox
- Printable design

We will examine the tools and techniques provided by HTML and CSS to develop elegant, mobile-first applications. By the end of this chapter, you will have a comprehensive understanding of how to create web applications that not only look great but also function flawlessly on any device, ensuring a consistent and user-friendly experience for all. We will also explore real-world examples where responsive design has made a significant impact, highlighting best practices and common pitfalls to avoid.

Technical requirements

The code files for this chapter can be found at `https://packt.link/YGV5J`.

What does it mean to design mobile first?

Designing mobile first is a good practice in developing digital products that prioritize the design and development of a website or application before scaling up to larger screens such as tablets and desktops.

This approach emerged as a response to the significant shift in how people access the internet. Historically, web design was centered on desktop computers, reflecting the dominant usage patterns of the early Internet era. However, with the advent of smartphones and their rapid adoption, particularly in the late 2000s and early 2010s, the landscape began to change dramatically. By 2016, mobile internet usage had overtaken desktops for the first time globally.

This shift necessitated a new approach: designing mobile first ensures that the most constrained environment is addressed first, ensuring a smooth, optimized experience for users on the go. Once the mobile experience is perfected, designers can enhance the interface for larger screens.

This method not only caters to the growing number of mobile users but also promotes cleaner, more efficient design practices, emphasizing essential content and functionality from the outset.

The mobile-first principle uses two key concepts:

- **Responsive web design (RWD)**: This involves a website adjusting its layout as the web browser is resized, ensuring it fits seamlessly on any device the user is utilizing.

- **Feature enhancements and graceful degradation**: This approach maximizes the capabilities of the device being used to access the website. For instance, on mobile websites, developers can leverage features such as GPS (Global Position System) to help users locate the nearest store, touch gestures for navigation (for example, swiping to see more content or pinching to zoom), and a gyroscope for interactive elements. On desktop websites, additional features can be incorporated to take advantage of the larger screen space while substituting touch gestures with mouse-click actions to maintain similar interactions. *Figure 8.1* shows the difference between a design that works on mobile from a real responsive design:

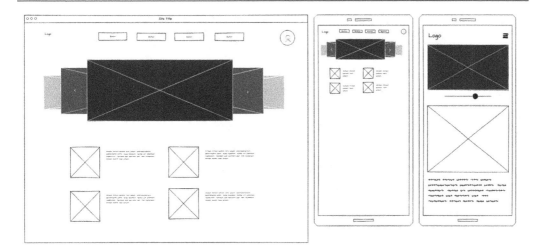

Figure 8.1: Wireframe for a desktop, a non-responsive design, and a responsive design, respectively

In *Figure 8.1*, you'll notice a stark contrast between a website displayed in its desktop layout, a non-responsive rendition, and a responsive version of the same content. The key to developing a robust responsive design lies in understanding the concept of the viewport in HTML, which we will discuss next.

Understanding viewports

In HTML, the **viewport** is essentially the visible area of a web page within a browser window. It determines how much content is visible and how it is scaled or sized to fit the screen. The viewport meta tag in HTML allows web developers to control the layout and scaling behavior of a web page on different devices and screen sizes.

Without specifying a viewport `meta` tag, mobile browsers typically render pages at a default desktop viewport width, which can lead to issues such as content appearing too small or requiring horizontal scrolling. The viewport meta tag is placed inside the `<head>` tag in HTML files:

```html
<!DOCTYPE html>
<html lang="en">
<head>
  <meta charset="UTF-8">
  <meta name="viewport" content="width=device-width,
    initial-scale=1.0">
  <title>Document</title>
</head>
<body>
```

```
</body>
</html>
```

By setting the viewport meta tag, developers can instruct the browser on how to adjust the page layout and scale it to fit the screen properly. Common attributes used in the viewport meta tag include `width`, `initial-scale`, `minimum-scale`, and `maximum-scale`, which allow developers to define the initial width, scale, and constraints for the viewport.

For example, specifying `<meta name="viewport" content="width=device-width, initial-scale=1.0">` instructs the browser to set the width of the viewport to the device's width and initially scale the content to 100% of its normal size.

Overall, understanding and properly configuring the viewport is crucial for ensuring a consistent and user-friendly experience across different devices and screen sizes in web development. The next concept we'll discuss, which is equally crucial for creating a good responsive design, is media queries.

Introducing media queries

Media queries were introduced as part of the CSS3 specification and gained significant popularity around the early 2010s as mobile devices with varying screen sizes became more prevalent, and there was a growing need for websites to adapt to different viewport sizes. They are a powerful tool used in RWD to apply different styles to a web page based on the characteristics of the device or browser viewing the page. These characteristics can include screen width, height, device orientation, resolution, and more. By utilizing media queries, developers can ensure their designs adapt seamlessly to different screen sizes and device capabilities, as illustrated in the following examples:

- Starting the code with the **default styles** for all sizes is a good practice that leverages the CSS cascade rule by ensuring default styles apply universally, with media queries overriding these defaults for specific conditions, thus creating a responsive and adaptable design:

```css
body {
  font-size: 16px;
  color: #333;
}

.sidebar {
  background-color: #a2cce4;
  padding: 20px;
  min-width: 150px;
}
.main-content {
  padding: 20px;
}
```

- Next, we look at a media query for screens smaller than 600px width:

```
@media screen and (max-width: 600px) {
  body {
      font-size: 14px;
  }
  .sidebar {
      display: none;
  }
}
```

This section of code will be displayed on viewports smaller than 600px, such as the smartphone shown in *Figure 8.2*:

Figure 8.2: Project on mobile screens

- The following is the media query for screens between 600px and 900px wide, with an adjustment for container width:

```
@media screen and (min-width: 600px) and (max-width: 900px) {
  .container {
    width: 90%;
  }
}
```

This media query defines the styles intended for viewports ranging from 600px to 900px in width, typical of tablets, as depicted in *Figure 8.3*:

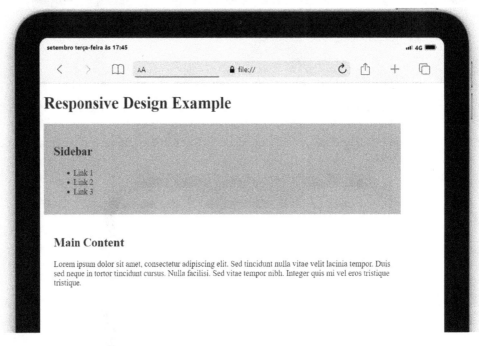

Figure 8.3: Project on tablets

- Next is the media query for screens larger than 1200px wide, with a limit for width on larger screens:

```
@media screen and (min-width: 1200px) {
  .container {
    width: 1200px;
    display: flex;
  }
}
```

- Finally, this code specifies the styles that will exclusively apply to desktops, as illustrated in *Figure 8.4*:

Figure 8.4: Project on desktops

In this example, the default styles are applied to all screen sizes. Also, the first media query targets screens smaller than 600px wide, adjusts the font size of the body, and hides the sidebar. The second media query applies styles to screens between 600px and 900px wide, adjusting the width of the container, while the third media query targets screens larger than 1200px wide and sets a maximum width for the container.

These media queries allow the web page layout and styling to adapt based on the viewport size, ensuring optimal display and user experience across various devices and screen resolutions.

Exercise 8.01 – Creating a mobile-first menu

In this exercise, we'll create a <nav> element representing a website's main menu. On desktop, it will be centralized in the header, but on mobile devices, it will be a vertical menu that can be further hidden using some JavaScript.

> **Important note**
>
> In this exercise, we'll be using CSS Flexbox, a concept that will be explained further in this chapter. Please feel free to jump to the *CSS Flexbox* section and then come back to complete this practice.

If you need help with this exercise, you can always check the solution code at this link: `https://packt.link/7IOiC`

Let's start our mobile-first menu:

1. First, create a folder named `chapter8-exercise1` and open it in VS Code.

2. Next, create basic HTML and CSS files in this folder. Name the HTML file `responsive-menu.html` and the CSS file `styles.css`.

3. Your folder structure should look like this in VS Code:

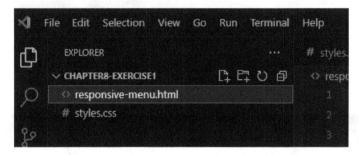

Figure 8.5: Project file structure example

4. Next, open the `responsive-menu.html` file and create the base HTML structure that we will style using the code provided here:

```html
<!DOCTYPE html>
<html lang="en">

<head>
  <meta charset="UTF-8">
  <meta name="viewport" content="width=device-width,
    initial-scale=1.0">
  <link rel="stylesheet" href="styles.css">
  <title>Responsive Menu</title>
</head>

<body>
  <header>
    <h1>Website Title</h1>
    <nav>
      <ul>
        <li>Home</li>
        <li>About Us</li>
        <li>Projects</li>
        <li>Partners</li>
        <li>Contact</li>
      </ul>
```

```
      </nav>
    </header>
  </body>

</html>
```

5. With the basic structure of our HTML ready, let's start by styling for mobile devices first. Open the `styles.css` file and write the code here, starting with the general styles:

```css
/* General Styles */
body {
  font-family: Arial, sans-serif;
  margin: 0;
  padding: 0;
  background-color: white;
  color: black;
}

header {
  background-color: purple;
  color: white;
  padding: 10px;
}

h1 {
  text-align: center;
  margin: 0;
  padding: 10px 0;
}
```

6. In the same `styles.css` file, let's add mobile-first styles in sequence:

```css
/* Mobile First Styles */
nav ul {
  list-style-type: none;
  padding: 0;
  margin: 0;
  display: flex;
  flex-direction: column;
  align-items: center;
}

nav ul li {
  padding: 10px;
  text-align: center;
  border: 2px solid grey;
  border-radius: 10px;
  margin: 5px 0;
```

```
    width: 80%;
    background-color: white;
    color: purple;
    transition: background-color 0.3s, color 0.3s;
}

nav ul li:hover {
    background-color: grey;
    color: white;
}
```

7. This CSS ensures that the website is styled effectively for mobile devices by default (vertical menu, centered items) and will later adjust for desktop views using media queries. It also includes visual enhancements such as rounded borders, color transitions, and hover effects for an improved user experience.

8. The application should look like this on both mobile and desktop devices:

Figure 8.6: Application functionality across mobile and desktop screens

9. Now, let's implement the desktop variation using media queries. For this, we'll use 820px as the minimum width for large-screen styling, as this works well for both tablets and desktops. In the styles.css file, below the previous code, add a media query for larger screens:

```
@media (min-width: 820px) {
    header {
```

```
        display: flex;
        flex-direction: column;
        align-items: center;
    }

    nav ul {
        flex-direction: row;
        justify-content: center;
    }

    nav ul li {
        margin: 0 10px;
        width: auto;
    }
}
```

These styles enhance the layout for larger screens, making the navigation menu horizontal and centering it, which is more suitable for desktop or tablet views. The result is as follows:

Figure 8.7: Application running on mobile, tablet, and desktop screens

Congratulations! You've successfully built your first mobile-first application. As you've experienced throughout this exercise, starting with a mobile-first approach simplifies the styling process, allowing for smoother transitions from simpler to more complex designs. Now, let's explore advanced concepts of responsive design to craft even more impressive applications!

Responsive values in CSS

Responsive values are fundamental for ensuring responsiveness on the web. By employing these values, developers can dynamically adjust layouts, typography, and other design elements to optimize user experiences across smartphones, tablets, desktops, and beyond. In this section, we'll delve into the principles and practical applications of responsive values in CSS, emphasizing their role in crafting flexible and user-friendly web interfaces.

Responsive typography units

Responsive typography units such as **root em** (rem) and em are important tools in web design for creating text that adjusts dynamically based on the user's preferred font size or the size of the parent element. They allow for scalability and consistency across different devices and screen sizes. Let's look at both of them in detail here:

- rem: The rem unit represents the font size of the root element (<html>), making it particularly useful for defining a consistent base font size for the entire document. One rem is equal to the computed font size of the root element. This means that if the root font size is 16px, 1rem would be equivalent to 16px.

- em: The em unit represents the font size of the current element relative to its parent element. For example, if the parent element has a font size of 16px and a child element is specified with 1em, it would have the same font size as the parent, 16px. If the parent's font size changes, the child element's font size will adjust proportionally.

Here's an example to illustrate the usage of rem and em:

```
html {
  font-size: 16px;
}
body {
  font-size: 1rem;
}
.container {
  font-size: 1.2rem;
}
.title {
  font-size: 2em;
}
.subtitle {
```

```
    font-size: 1.5em;
}
```

In this example, we have the following:

- The `html` element sets the base font size for the document to `16px`
- The `body` element sets its font size to `1rem`, which is `16px` (the size of the root element)
- The `.container` class sets its font size to `1.2rem`, which is `20.8px` (`16px * 1.2`)
- The `.title` class sets its font size to `2em`, which is `32px` (`16px * 2`)
- The `.subtitle` class sets its font size to `1.5em`, which is `24px` (`16px * 1.5`)

Using `rem` and em units allows for flexible and scalable typography that adapts to changes in the document structure and makes it easier to maintain consistent typography across different parts of a website.

In sequence, we'll learn about responsive sizing units, such as **viewport height** (vh) and **viewport width** (vw).

Responsive sizing units

Responsive sizing units vh and vw are used in CSS to specify dimensions relative to the viewport's height (vh) and width (vw). These units are particularly useful for creating layouts that adapt to different screen sizes and orientations, providing a responsive design experience:

- vh: This unit represents `1%` of the viewport's height. For example, `100vh` would be equal to the full height of the viewport, while `50vh` would be half of the viewport's height. It's commonly used for elements that need to fill the entire height of the viewport, such as full-screen sections or backgrounds.
- vw: Similar to vh, vw represents 1% of the viewport's width. For instance, `100vw` would be the full width of the viewport, while `50vw` would be half of the viewport's width. It's often used for creating responsive layouts or sizing elements relative to the width of the viewport.

Here's an example to illustrate the usage of vh and vw:

```
.container {
  width: 80vw;
  height: 70vh
  background-color: #f0f0f0;
}
.fullscreen-section {
  height: 100vh
  background-image: url('background.jpg');
  background-size: cover;
}
```

In this example, we have the following:

- The `.container` class sets the width to `80%` of the viewport width (`80vw`) and the height to `70%` of the viewport height (`70vh`)

- The `.fullscreen-section` class ensures that the section fills the entire height of the viewport (`100vh`), creating a full-screen background image

Using `vh` and `vw` units allows developers to create more flexible and adaptive layouts that adjust smoothly to different screen sizes, making them essential tools for RWD. In the following section, we will explore another crucial aspect of responsive design: how to create responsive images.

Exploring responsive images

Images play a crucial role in developing different viewport sizes. Beyond using CSS, HTML also provides resources that enhance the user experience in responsive applications. Let's explore these tools.

Using width and max-width with percentages

While `width` and `max-width` are not exclusively dedicated to responsive design, their effective use with percentages significantly contributes to achieving responsive images. Here's how they function:

- Setting the image `width` value to `100%` ensures that the image scales with its parent container:

```
<style>
  .responsive-img {
    width: 100%;
    height: auto;
  }
</style>
<img src="example.jpg" alt="Example Image"
  class="responsive-img">
```

- Setting `max-width` to `100%` ensures that the image will not exceed the size of its container while maintaining its aspect ratio:

```
<style>
  .responsive-img {
    max-width: 100%;
    height: auto;
  }
</style>
<img src="example.jpg" alt="Example Image"
  class="responsive-img">
```

While the `width` property determines the **exact width** of an element, the `max-width` property sets the **maximum width** that an element can have. This allows elements to be flexible and adapt to various screen sizes or content lengths without surpassing the specified maximum width.

The next resource for responsive images is the HTML `picture` element.

HTML picture element

The `picture` element allows you to define multiple sources for an image, so different images can be served based on the viewport width. Here's how you might use this element:

```
<picture>
    <source srcset="image-large.jpg"
            media="(min-width: 800px)">
    <source srcset="image-medium.jpg"
            media="(min-width: 500px)">
    <img src="image-small.jpg"
        alt="Responsive Image">
</picture>
```

To summarize, the `picture` attribute helps ensure image quality and prevents distortion.

Using srcset and sizes

The `srcset` attribute allows you to specify different image resolutions, and the `sizes` attribute lets the browser know what image size to select based on the viewport width:

```
<img src="example-small.jpg"
    srcset="example-small.jpg 300w,
            example-medium.jpg 600w,
            example-large.jpg 1200w"
    sizes="(max-width: 600px) 300px,
            (max-width: 900px) 600px,
            1200px"
    alt="Example Image">
```

The `srcset` attribute allows browsers to choose the most appropriate image source based on factors such as device resolution and viewport size, thus optimizing performance and displaying the most suitable image for each user.

Sometimes, we use images as backgrounds, and CSS provides specific properties to handle their responsiveness, as we'll discuss next.

Background images in CSS

For images used as backgrounds, the `background-size` property can make them responsive:

```
<style>
  .responsive-bg {
    background-image: url('example.jpg');
    background-size: cover;
    background-position: center;
    width: 100%;
    height: 400px;
  }
</style>
<div class="responsive-bg"></div>
```

The `background-size` property is commonly used in conjunction with the `background-image` property to control the appearance of background images in elements. It is useful for creating responsive designs and adjusting the presentation of background images based on the dimensions of the container or the viewport. The `background-size` property accepts the following values:

- `cover`: This value scales the background image as large as possible without stretching it while ensuring that both dimensions are fully covered by the background positioning area. This means that the background image might be cropped in order to fit the container's dimensions.

- `contain`: This value scales the background image to make sure it is fully contained within the background positioning area without cropping or stretching it. This means that the entire background image will be **visible**, even if it leaves some empty space within the container:

Figure 8.8: Image with background-size cover and contain values, respectively

Media queries can also change the background image based on the viewport size:

```
<style>
  .responsive-bg {
    background-image: url('example-small.jpg');
    background-size: cover;
    background-position: center;
  }

  @media (min-width: 600px) {
    .responsive-bg {
      background-image: url('example-medium.jpg');
    }
  }

  @media (min-width: 1200px) {
    .responsive-bg {
      background-image: url('example-large.jpg');
    }
  }
</style>
<div class="responsive-bg"></div>
```

In this code, we assign a different background image for each viewport size. This strategy ensures optimal image quality and positioning while reducing rendering costs, as larger images are only displayed on larger viewports.

When dealing with images, one of the first questions that developers often ask is: How long will it take to render? As we explored earlier, modern CSS and HTML include special properties designed to enhance both image quality and performance. Next, we'll examine another strategy aimed at reducing image loading times: lazy loading.

Lazy loading

Lazy loading is a strategy to load content on demand in an application. It improves performance by only loading what is visible on the screen, with additional content loading as the user scrolls near it.

In HTML, we can use the `loading` attribute in images to defer offscreen images until the user scrolls near them, further enhancing performance:

```
<img src="example.jpg" alt="Example Image" loading="lazy">
```

Responsive images are crucial for ensuring that web content looks good on various devices and screen sizes. The techniques discussed in this section collectively ensure that images adapt seamlessly to different viewports, significantly enhancing the overall user experience.

For the final part of this section, let us look at an exercise.

Exercise 8.02 – Creating a mobile catalog for a flower shop

Let's enhance our skills by applying what we've learned so far. We're embarking on a journey to design a captivating home page for a flower shop. To kickstart this process, we'll build upon the foundation laid in *Exercise 8.01*, enriching our home page with additional content and enhancements:

1. Begin by creating a new directory named `chapter8-exercise2` and duplicating the files from *Exercise 8.01* into this new directory. These files will serve as the groundwork for our refinement and expansion.

2. Open the directory using VS Code and rename the HTML file from `responsive-menu.html` to `responsive-shop.html`.

3. Next, access the provided assets located in the `chapter8-exercise2` folder for this exercise. Within this folder, you'll also discover the resolution code for reference: `https://packt.link/axx5J`.

4. Ensure your folder structure resembles the following:

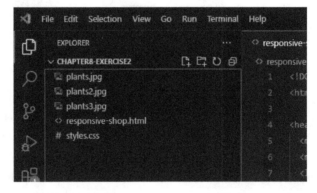

Figure 8.9: Project folder structure

5. To enrich our flower shop's home page, we'll introduce a new section showcasing our products. Begin by opening the `responsive-shop.html` file and inserting the following code snippet immediately after the closing `</header>` tag:

```
<div class="container">
    <div class="card">
      <img src="./plants.jpg" alt="Plant 1">
      <h2>Responsive Content</h2>
    </div>
    <div class="card">
      <img src="./plants2.jpg" alt="Plant 2">
      <h2>Responsive Content</h2>
```

```
      </div>
      <div class="card">
        <img src="./plants3.jpg" alt="Plant 3">
        <h2>Responsive Content</h2>
      </div>
    </div>
  </div>
```

6. This code snippet represents the new product section for our flower shop. After integrating this content, your `responsive-shop.html` file should resemble the same as the example found at this link: `https://packt.link/pt5gS`.

7. You've set up the HTML file, but upon inspection, you may notice that the images appear too large and lack the appearance of product cards. Let's enhance the visual presentation by styling the new elements to create an appealing mobile catalog tailored for smaller screens. Additionally, we'll transition our measurement values to responsive units.

8. Firstly, it's good practice to apply a CSS reset to neutralize any default browser styling. Open the `styles.css` file and insert the following snippet just before any existing code:

```
* {
  box-sizing: border-box;
  margin: 0;
  padding: 0;
}
```

9. Next, as we're employing relative sizes for fonts and elements, let's employ a technique to simplify the transition from pixels (px) to rems (rem). Typically, `1rem` equals `16px`. By reducing the default base font size of the browser by `62.5%`, we can establish `1rem` as `10px`, making calculations more straightforward. Add the following line of CSS immediately after the reset:

```
:root {
  font-size: 62.5%
}
```

10. We also need to style the new `.container`, `.card`, and image (`img`) elements. Note that the `.container` element is currently set to a width of `100vw`, meaning it will span the entire viewport width:

```
.container {
  margin: 0 auto;
  padding: 2rem;
  max-width: 100vw;
}
.card {
```

```
      margin-bottom: 2rem;
      padding: 2rem;
      background-color: rgb(224, 203, 224);
  }
  img {
    width: 100%;
  }
```

11. You'll notice that these new elements are already using relative units (rem) for sizing. Now, refactor the remaining properties in your CSS file to also use `rem` units. You can find the resulting CSS file at this link: `https://packt.link/JUuqH`.

The result is as follows:

Figure 8.10: Visual output of the code

Great job! The flower shop's catalog is now elegantly designed for mobile devices. Take the opportunity to delve into CSS resources for animations and additional styling options according to your preferences. Next, let's delve into the powerful Flexbox module to enhance our application effortlessly for desktops and larger screens.

CSS Flexbox

CSS Flexbox is a powerful layout module designed to help developers create flexible and responsive web layouts with ease. It allows items within a container to be automatically arranged and adjusted based on the available space, making it particularly useful for responsive design. Here are the key concepts of Flexbox:

- **Flex container and flex items**: The parent element becomes a flex container when `display: flex` or `display: inline-flex` is applied. The children of this container become flex items and are arranged according to Flexbox rules.

- **Main axis and cross axis**: The main axis is the primary direction in which flex items are laid out (default is horizontal). The cross-axis is perpendicular to the main axis.

Figure 8.11 shows how flex items relate to a flex container:

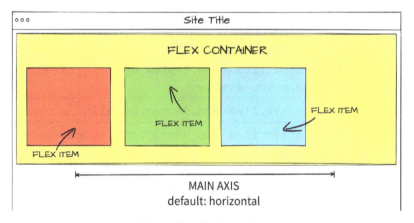

Figure 8.11: Flexbox schema

To make the most of CSS Flexbox, let's delve deeper into its properties and explore how to use them effectively.

Flexbox basic properties

As the `display: flex` property is set in the parent element, it gives us a set of properties to better configure the element positioning and its children. Let's take a look at each one of them:

- `display: flex` is used to declare a flex container; usually, it is the first we set
- `flex-direction` is the property that determines the direction of flex items
- `justify-content` is used to align items along the main axis inside the flex container
- `align-items` also is used to align items, but using the cross axis as a reference
- `flex-wrap` is used to control whether flex items wrap
- `align-content` is a property used to align each flex container line if there's any extra space in the cross-axis

Each flex container property accepts different values, as shown in *Table 8.1*:

Property	Accepted Values
`display`	`flex, inline-flex`
`flex-direction`	`row, row-reverse, column, column-reverse`
`flex-wrap`	`nowrap, wrap, wrap-reverse`
`justify-content`	`flex-start, flex-end, center, space-between, space-around, space-evenly`
`align-items`	`flex-start, flex-end, center, baseline, stretch`
`align-content`	`flex-start, flex-end, center, space-between, space-around, stretch`

Table 8.1: Flex container properties and values accepted

The main axis is horizontal by default because the default value for `flex-direction` is `row`. If `flex-direction` is changed to `column`, the main axis is automatically set to vertical. The **flex item** properties are the following:

- `order`: Determines the sequence of flex items
- `flex-grow`: Defines how much a flex item can expand in relation to the other items
- `flex-shrink`: Determines how much a flex item can reduce its size compared to others

- `flex-basis`: Sets the initial size of an element before space is allocated
- `align-self`: Lets individual flex items override the default alignment

Each flex item property accepts different values, as shown in *Table 8.2*:

Property	Accepted Values
`flex-grow`	Any positive number, 0
`flex-shrink`	Any positive number, 0
`flex-basis`	Any length value, percentage, `auto`
`flex`	Shorthand for `flex-grow`, `flex-shrink`, and `flex-basis`
`order`	Any integer
`align-self`	`auto`, `flex-start`, `flex-end`, `center`, `baseline`, `stretch`

Table 8.2: Flex item properties and values accepted

Please refer to the following link to see the full code demonstrating how Flexbox properties can simplify responsive styling: `https://packt.link/rmaSc`

In this code, the `.container` class is set to `display: flex` with `flex-wrap: wrap` to allow wrapping of flex items, styled as follows:

```
.container {
    display: flex;
    flex-wrap: wrap;
    justify-content: space-between;
}
```

The items are styled as follows:

```
.item {
    flex-grow: 1;
    flex-shrink: 1;
    flex-basis: 200px;
    margin: 10px;
    padding: 20px;
    background-color: #f0f0f0;
    text-align: center;
}
```

```
.item:nth-child(even) {
    background-color: #e0e0e0;
}
```

In this code, the `.item` class is styled as follows:

- `flex-grow: 1`: Each item can grow to fill the available space
- `flex-shrink: 1`: Each item can shrink if necessary
- `flex-basis: 200px`: Each item has a base size of 200px

To ensure responsiveness, the media query guarantees that on screens smaller than 600px, each flex item takes up the full width (100%), as shown in the following code:

```
@media (max-width: 600px) {
    .item {
        flex-basis: 100%;
    }
}
```

In the example, we see that while Flexbox is powerful, it's also important to use other tools provided by HTML and CSS, such as media queries, to achieve optimal responsiveness. The result of the code is shown in *Figure 8.12*:

Figure 8.12: Flexbox in desktop

When resized, the code automatically reorganizes itself, as shown in *Figure 8.13*:

Figure 8.13: Flexbox in tablet and mobile

This example demonstrates how to use Flexbox to create a responsive layout that adapts to different screen sizes, providing a flexible and robust solution for web development.

Next, let's look at an exercise that uses Flexbox.

Exercise 8.03 – Updating the flower shop catalog for desktop using Flexbox

Let's create the desktop version of the flower shop catalog we developed in *Exercise 8.02* using Flexbox. Begin by establishing a new directory named `chapter8-exercise3`.

If you have any queries or require guidance, refer to the resolution code provided for this exercise in the project folder: `https://packt.link/jTw80`

This time, we'll focus exclusively on leveraging CSS Flexbox for layout optimization without altering the HTML file. However, feel free to enhance the project with additional resources you've learned throughout this book:

1. After duplicating the files from *Exercise 8.02* into the new folder, ensure that your file structure appears as follows in VS Code:

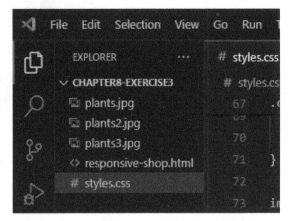

Figure 8.14: Exercise 8.3 folder structure

2. Next, navigate to the `styles.css` file and locate the media query designated for screens larger than `820px`. Within this query, we'll convert the `.container` class into a Flexbox container, adjusting its positioning and its child elements accordingly:

```css
/* Bigger Screens Styles */
@media (min-width: 820px) {
  header {
    display: flex;
    flex-direction: column;
    align-items: center;
  }

  nav ul {
    flex-direction: row;
    justify-content: center;
  }

  nav ul li {
    margin: 0 1rem;
    width: auto;
  }
  .container {
```

```
        display: flex;
        justify-content: space-around;
        gap: 2rem;
        max-width: 120rem;
    }
    img {
        width: auto;
        max-height: 27.3rem;
    }
}
```

3. The gap property is permitted in Flexbox containers and specifies the spacing between flex items.

4. We've additionally applied a max-height value to the images to ensure they fit more appropriately within the cards.

5. Furthermore, you may observe that we replaced width: 100% with width: auto to negate the previously set width value. This adjustment is necessary due to the cascade rule of CSS, which stipulates that if a property is not explicitly altered within the most specific criteria (in this instance, the media query), it will inherit the global value set beforehand.

The output for our flower shop is as follows:

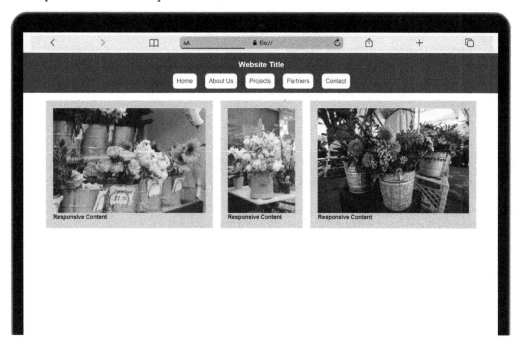

Figure 8.15: Visual output of the exercise

With just a few lines of code, we've transformed the content into a more appealing design for larger screens! Incredible! However, there are occasions when users may need to print content for various reasons, prompting developers to utilize responsive tools to enhance the print experience. This will be our focus in the next section.

Exploring printable design

Responsive design traditionally refers to making web content adaptable to various screen sizes and devices, ensuring a seamless user experience across different viewports. However, the core principle of responsive design is to adapt content to different contexts and environments, which includes printing.

The following list shows how printable design is a form of responsive design:

- **Adaptability**: Just as responsive design adjusts layouts for different screen sizes, printable design adjusts the web page layout for optimal printing. This includes changes in font sizes, colors, margins, and visibility of elements.

- **Media queries**: Both responsive design for screens and printable design use CSS media queries to apply different styles based on the medium (for example, screen or print).

- **User experience**: The goal in both cases is to improve user experience. For RWD, this means making the content easy to read and navigate on various devices. For printable design, this means making the content clean, well structured, and easy to read on paper.

There are two ways to specify print-only styles in a browser:

- **Using a separate style sheet**: You can import a separate style sheet dedicated to print styles with a new HTML `<link>` tag:

```
<link rel="stylesheet"
      media="print"
      href="print.css" />
```

This code allows the `print.css` style sheet to be used only when the browser is in print mode.

- **Using @media queries in an existing style sheet**: You can include print-specific styles within an existing style sheet by using the `@media print` query syntax:

```
@media print {
  /* Print-specific style changes go here */
}
```

Both methods are effective for applying styles, specifically when a document is printed. Using a separate style sheet helps keep your print styles organized and distinct from your screen styles, while using `@media print` within an existing style sheet allows you to maintain all styles in a single file, which can be more convenient for smaller projects or simpler style adjustments.

At the following link, `https://packt.link/bu7NR`, is an example demonstrating how to create a web page with a separate print style sheet and how to use `@media print` within an existing style sheet to achieve printable responsiveness. In this code, the HTML includes two style sheets: `styles.css` for general styles and `print.css` for print-specific styles.

Let's look at the styles now. The main CSS file (`styles.css`) defines basic styles for the web page, ensuring it looks good on screens, and uses media queries to adjust styles for smaller screens, demonstrating responsive design principles. The main style sheet is available at this link for you to study: `https://packt.link/acNvH`. *Figure 8.16* represents the visual output for this code:

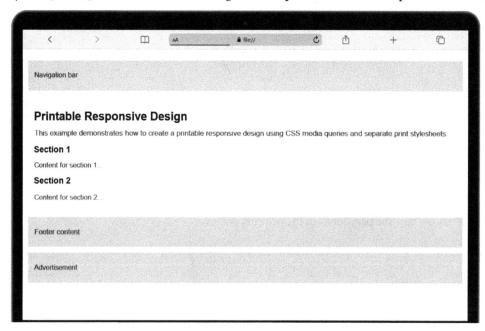

Figure 8.16: Screen view before printing

The print CSS file (`print.css`) applies styles specifically for when the page is printed, adjusts body margins, font size, and background colors for a clean print layout, and hides elements such as navigation elements, footers, and ads that are not needed in the printed version.

This CSS file also ensures headings avoid page breaks and sections are printed on separate pages for better readability. The result is as follows:

```
@media print {
  body {
```

```
    margin: 1in;
    font-size: 12pt;
    color: black;
    background-color: white;
}

.navigation, .footer, .ad {
    display: none;
}

.content {
    padding: 0;
}

h1, h2, h3 {
    page-break-before: avoid;
}

.section {
    page-break-after: always;
}
}
```

Figure 8.17 shows how the page looks in the printing preview:

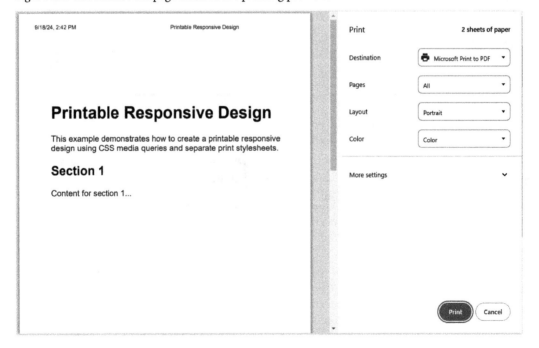

Figure 8.17: Printing preview

In conclusion, mastering print styles is an essential aspect of web development that enhances the versatility and professionalism of your web content. This involves hiding unnecessary elements, adjusting layouts, and managing page breaks to create clean and readable printed documents.

Combining these techniques with a solid understanding of Flexbox and responsive design principles ensures that your web content is adaptable and user-friendly across all devices and formats, ultimately improving the overall user experience.

The final part of this chapter is an exercise for creating printable designs.

Exercise 8.04 – Creating a printable style for the flower shop catalog

Let's enhance our skills by optimizing our web page for printing. We'll be working with the HTML and CSS files provided to create a print-friendly version of our flower shop catalog:

1. Start by creating a new directory named `chapter8-exercise4`. Duplicate the files from the previous exercise (*Exercise 8.03*) into this new directory. These files will serve as the foundation for our print optimization practice. Remember to look at our GitHub repository to check the assets and resulting code: `https://packt.link/f7grZ`.

2. Open the directory in VS Code. In this exercise, we'll focus on creating CSS styles specifically tailored for printing purposes, ensuring that the printed version of our catalog looks clean and well structured. Create a new file and name it `print.css`. The structure of your project should be like this:

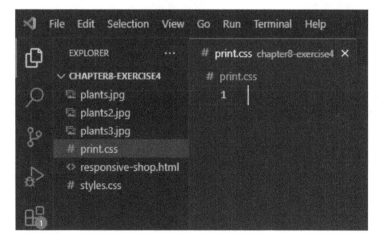

Figure 8.18: Project file structure

3. Open the `print.css` file and include the code as shown here:

```
/* Printing Styles */

@media print {
body {
  font-family: Arial, sans-serif;
}
header {
  display: none;
}
.container {
  display: block;
  margin: 0 auto;
}
.card {
  page-break-inside: avoid;
  margin-bottom: 2rem;
  padding: 2rem;
  background-color: rgb(224, 203, 224);
}
img {
  display: block;
  width: 100%;
  max-height: 100%;
}
nav {
  display: none;
}
}
```

4. Let's break down the CSS code for printing styles:

- body: We keep the font family as Arial or sans-serif for better readability in print.

- header: We set the header to be hidden (display: none;) in print mode, as headers are typically not needed in printed documents.

- .container: We ensure that the container appears as a block-level element (display: block;) and is centered horizontally (margin: 0 auto;) on the printed page.

- .card: We use page-break-inside: avoid to prevent splitting the cards across pages, ensuring that each card stays together on one page for better readability.

- img: We set the images to be displayed as block-level elements (display: block;), ensuring that they take up the full width of their container. Additionally, we limit the maximum height of images to prevent them from exceeding the height of the printed page (max-height: 100%;).

- nav: We hide the navigation menu in print mode by setting `display: none`, as navigation menus are typically not needed in printed documents.

5. These CSS rules help optimize the page layout and content for printing, ensuring that the printed document looks clean and well-structured.

6. Next, let's integrate the newly created printing styles into the HTML file by adding a new `<link>` tag below the main style sheet:

```
<head>
  <meta charset="UTF-8">
  <meta name="viewport" content="width=device-width,
    initial-scale=1.0">
  <link rel="stylesheet" href="styles.css">
  <link rel="stylesheet" href="print.css">
  <title>Responsive Menu</title>
</head>
```

The result is as follows:

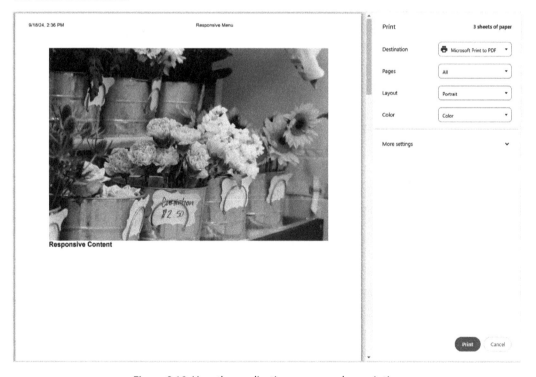

Figure 8.19: How the application appears when printing

By optimizing our web page for printing, we've acquired a valuable skill set that enhances user experience beyond the digital realm. This exercise underscores the importance of catering to diverse user needs and reinforces our commitment to delivering exceptional usability across all platforms.

Summary

Through this chapter, we've explored various techniques for creating responsive designs. We began with the concept of mobile-first development and its importance. We then covered **viewports**, **media queries**, and responsive measurement units such as rem, em, vh, and vw. Additionally, we delved into the **Flexbox** layout and how to leverage it to simplify responsive design. Finally, we learned how to implement techniques to make **printed** content more user-friendly, enhancing the overall experience of using our applications. With this knowledge, we hope you'll create incredible applications that run seamlessly on any screen size, providing an excellent user experience.

In the next chapter, we'll delve deeper into user experience features, focusing on HTML and CSS accessibility tools. This will enable you to develop fully accessible applications that offer an outstanding user experience for everyone, regardless of their conditions.

9

Ensuring Accessibility in HTML and CSS

Web accessibility is a broad set of guidelines, considerations, and best practices that we can apply to make sure we don't put up barriers that stop those with disabilities from using web pages.

The web is an open platform, and we should strive to make its content accessible to as many people as we can. The web should work for everyone. As a web developer, this means we have a great responsibility to use the tools available to us to make accessible websites.

HTML has a lot of features that are designed to make web pages accessible and, in this chapter, we will look at several of these – image `alt` attributes, form labels, and fieldsets, to name a few.

We will start by identifying some common issues that can cause web pages to become inaccessible. We will look at how a combination of thinking with an accessibility mindset from the outset of our project and using semantic HTML as intended will solve a lot of these issues.

We will also look at some of the tools that are available to help us improve the accessibility of our web pages. We will be looking at a tool called Axe that can automatically flag accessibility issues.

Finally, as an activity, we will look at a page with some accessibility issues and use what we have learned in this chapter to identify and fix those issues.

In this chapter, we're going to cover the following main topics:

- Making images accessible
- Making forms accessible
- Working with accessibility tools

By the end of this chapter, you will understand how to make an accessible web page and how to discover accessibility issues using readily available tools. You will also have created an accessible web page with images and forms that can be used by disabled users.

Technical requirements

The code files for this chapter can be found at `https://packt.link/4oGcq`.

What is accessibility?

Accessibility is a very important subject for the web and web developers, but it is one that isn't always that well understood. By learning about some of the fundamentals of web accessibility, we can make small improvements that will make a big difference for a lot of users.

> **Note**
>
> Accessibility is a long word often shortened to the numeronym (a word abbreviated with numbers) a11y. The term a11y is a useful one to recognize as it is widely used when talking about accessibility and, if you are researching the subject, a web search for a11y will return useful results.

We can remove barriers for the following users:

- Users with visual impairments who cannot get information from images that do not have a text alternative

- Users with hearing impairments who cannot get information from media (audio or video)

- Users with physical impairments that prevent them from using a mouse

By making our web pages accessible to those who are disabled, we also make the pages more usable for those who may face technical limitations. Accessibility issues are not only limited to the aforementioned scenarios but also include users on mobile devices, users who are on a website in a location with a lot of background noise, during a presentation, or on a monitor that is not well calibrated, and where color contrast may be an issue. Simply put, accessibility improves usability.

As web developers, one of the biggest opportunities we have for creating accessible web pages is the use of semantically correct HTML. What this means is that we should make use of the correct HTML tags to define the meaning of the content of our page.

For example, instead of using `div` elements to create the sections of a page, we should use the elements provided by HTML5 – `section`, `header`, `footer`, `main`, `aside`, and `nav`. Also, we should use `h1-h6` and use the right level of heading, depending on our document structure and nesting.

The web is a great platform for distributing content, and the main purpose of that content is to reach as many people as possible. Whether our reasons are due to ethics, profit margins, fear of litigation, or empathy, there are no good reasons to ignore the accessibility needs of our users.

You'll be happy to hear that we've already learned about a lot of the techniques we need to make our web pages accessible. Before we progress, let's review some of the accessibility concepts we have covered in the preceding chapters and how we will be building upon them in this chapter.

In *Chapter 1*, we talked briefly about the `alt` attribute on the `img` tag. In the next section of this chapter, we will look at how we can make images accessible to visually impaired users (and those with image loading disabled) using the `alt` attribute in the right way.

The `alt` attribute is often misused, but by making some simple, informed decisions we can more easily pick the right approach for `alt` text depending on the context in which an image is used and the information it is being used to convey.

In *Chapter 4*, we looked at creating HTML forms. We will reinforce some of that learning by focusing on the fundamentals for making these forms accessible, including connecting labels to inputs correctly and keyboard accessibility.

An area that we have not covered previously and one that is specific to enhancing the accessibility of web pages is the **Web Accessibility Initiative's** suite of standards known as **ARIA** (*Accessible Rich Internet Applications*).

The WAI-ARIA standards are extensive and are often used to enhance web pages with a lot of dynamism. Put simply, where a web application uses JavaScript to update a page often, WAI-ARIA can help maintain the meaning of the page, keep a user's focus in the right place, and keep behavior accessible. The ARIA attributes and roles are beyond the scope of this chapter as they are largely related to JavaScript and interpreting custom behaviors as something a screen reader can understand.

First, we will look at how we can make our HTML images accessible.

Making images accessible

As the saying goes, a picture is worth a thousand words, and images can add a lot to a web page. Some pictures are decorative, whereas others are an important piece of content that gets your web page's message across with great impact.

Every time we add an image to a web page, there are accessibility considerations we should take into consideration in order to make the image accessible. Not all users may be able to see an image. For example, if a user requires a screen reader or another form of non-visual browser to be able to navigate through a web page, they will require a textual description of an image to be able to garner any meaning from it. It can also be the case that a user does not download images because of limitations on network bandwidth or for security reasons. All of these users will benefit from an alternative text description of an image.

The way we can provide this information is through the `alt` attribute, which we learned about in *Chapter 1*. An `alt` attribute should provide a meaningful text alternative to an image.

For example, if we were to add an image to our web page that showed an infographic with some important business data, such as the budget for this and last year, the information – any text, labels, or the numbers the bars represent – would not be accessible to a non-visual browser if they were included in the image. We would need to add an `alt` tag that expressed that information.

The following `img` tag could be used to provide an image and appropriate `alt` text:

```
<img
  src="bar-chart.png"
  alt="Bar graph of profits for 2019 (£40,000), which are
       up £20,000 on profits for 2018 (£20,000)"
/>
```

The following is our infographic:

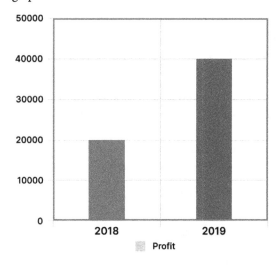

Figure 9.1: Bar graph infographic

The following screenshot shows the `alt` tag (with the image not loaded). The `alt` tag can also be read by a screen reader:

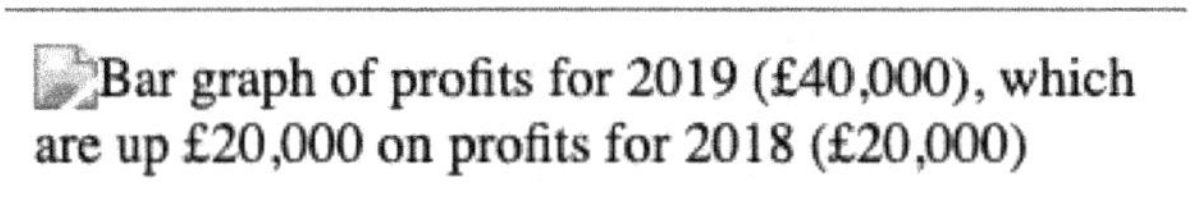

Bar graph of profits for 2019 (£40,000), which are up £20,000 on profits for 2018 (£20,000)

Figure 9.2: alt text for the bar graph image

There are a few decisions we should consider when we add an `alt` attribute, and these depend on the nature of the image (or images) on our web page.

> **Note**
>
> The W3C provides a decision tree that can be very useful for making a decision about what content you need to provide in the `alt` attribute of an image on your web page. You can find it at `https://www.w3.org/WAI/tutorials/images/decision-tree/`.

As we have shown in the previous example, informative images need a text alternative. Where an image provides the user with information in the form of a diagram, graph, photo, or illustration that represents data or a concept, there should also be an `alt` text attribute. We should, at least, provide a short text description of the essential information the image conveys. Often, and this especially applies where the information in the image is complex, we need to back this up with more text content or with a table of data.

If an image is purely decorative, we should still provide an `alt` attribute, but we don't need to add any text. We just add an `alt` attribute with an empty string (`alt=""`). Decorative images include those that are used for visual styling and effects.

Images that are described by surrounding text on the page can also be considered decorative. If the information that's conveyed by an image is also described as text on a web page, you can again add an empty `alt` tag.

If an image has a functional role – for example, it is an icon that's used as a button or a link, we need to provide alternative text that describes the function rather than the image. For example, where an image of a floppy disk is used as a button for saving a file, we would provide the `alt` text *Save* or *Save file* rather than something such as *Floppy disk icon*, which would describe the image but would not be useful for a user who wanted to use that functionality.

Where multiple images are used to convey a single piece of information, we can add `alt` text to the first image to provide a description. The rest of the images can then have an empty `alt` attribute. For example, in the forthcoming exercise, we will create an element for showing a rating using five separate star images. By adding an `alt` text description to the first image and empty `alt` attributes to the other images, we inform the user of the rating, even when the images are not visible.

Here is the relevant code:

```
<div class="rating">
  <img
    src="images/full-star.png"
    alt="Rated 3 and a half out of 5 stars"
  />
  <img src="images/full-star.png" alt="" />
  <img src="images/full-star.png" alt="" />
  <img src="images/half-star.png" alt="" />
  <img src="images/empty-star.png" alt="" />
</div>
```

Having seen the basics of the `alt` attribute, we will implement this in the following exercise, where we will make the rating section of a typical product page accessible. We will make the group of images accessible with the appropriate use of `alt` text.

Exercise 9.01 - Accessible ratings

In this exercise, we are going to create a product page. We are going to include an element for showing a product rating. This is a common UI pattern seen on e-commerce sites or anywhere users can rate cultural artifacts such as books and films. The rating will be represented by five stars, and a rating can be any value from zero to five stars, rising in increments of a half.

To create the rating element, we will use five separate images. We will make the group of images accessible by providing `alt` attributes to express the information on our set of images. In other words, we will describe the rating given.

The following screenshot is what the product page will look like when it's finished. For this exercise, we will give the product a rating of three and a half stars:

Product Description

Lorem ipsum dolor sit amet, consectetur adipiscing elit. Phasellus scelerisque, sapien at tincidunt finibus, mauris purus tincidunt orci, non cursus lorem lectus ac urna. Ut non porttitor nisl. Morbi id nisi eros.

Figure 9.3: Product page with rating UI

The steps are as follows:

1. First, we will create an empty web page in which to build up our accessible product page. We will create and save a file named `Exercise 9.01.html` and add the following code to provide a page title and to add some styles to remove white space from the body element:

```html
<!DOCTYPE html>
<html lang="en">
    <head>
        <meta charset="utf-8">
        <title>Exercise 9.01:Accessible rating</title>
        <style>
        html {
            box-sizing: border-box;
        }

        *, *:before, *:after {
            box-sizing: inherit;
        }

        body {
            padding: 0;
            margin: 0;
            font-family: Arial, Helvetica, sans-serif;
            font-size: 16px;
            line-height: 1.7778;
        }
        </style>
    </head>
    <body>

    </body>
</html>
```

2. In the body element, add a `<div>` with a `class` attribute with the `container` value. This is where our product markup will be hosted:

```html
<div class="container"></div>
```

3. We want the product to be centered on the page, so we will use flexbox to center the contents of the container in CSS. By setting the flexbox flow to a column with no wrapping, the elements within that container will flow vertically, one after another. We add the following declaration block to the `style` element:

```css
.container {
    display: flex;
    align-items: center;
```

```
        justify-content: center;
        flex-flow: column nowrap;
}
```

4. We are going to create a product item page that shows an item with a title, a description, and a rating. We can add the initial markup for the product – a section element with the `product` class attribute and child elements for a product image, as well as a heading and a paragraph for the description. We add this to the body element:

```
<section class="product">
    <h2 class="product-heading">
        Product Description
    </h2>
    <img
        class="product-image"
        src="images/product.png"
        alt="Product"
    />
    <p class="product-description">
        Lorem ipsum dolor sit amet,
        consectetur adipiscing elit.
        Phasellus scelerisque, sapien at
        tincidunt finibus, mauris purus
        tincidunt orci, non cursus lorem
        lectus ac urna. Ut non porttitor
        nisl. Morbi id nisi eros.
    </p>
    <hr class="divider" />
</section>
```

5. To style the product item, we will add the following CSS declarations. This defines a box for the product with a shadow around it and the content centered:

```
.product {
    width: 50vw;
    min-width: 640px;
    margin: 2rem;
    padding: 1rem 2rem;
    display: flex;
    justify-content: center;
    flex-flow: column nowrap;
    border-radius: 3px;
    box-shadow:
        rgba(0, 23, 74, 0.05) 0px 6px 12px
            12px,
        rgba(0, 23, 24, 0.1) 0 6px 6px 6px,
```

```
            rgba(0, 23, 24, 0.3) 0 1px 0px 0px;
}

.product-heading {
    margin: 0;
}

.product-image {
    padding: 0;
    margin: 2rem auto;
    width: 60%;
    height: 100%;
}

.product-description {
    width: 100%;
}

.divider {
    width: 100%;
}
```

6. To create the UI element for showing the ratings, we will add a `div` element with a `rating` class attribute and five image elements as children. Each of these images will either show a star, a half-filled star, or an empty star. We will add this below the description at the end of the product section:

```
<div class="rating">
    <img src="images/full-star.png" />
    <img src="images/full-star.png" />
    <img src="images/full-star.png" />
     <img src="images/half-star.png" />
    <img src="images/empty-star.png" />
</div>
```

7. We need to add some more CSS declarations to the style element to center the ratings and to keep the size of the rating responsive when the container is resized due to window width. We have added `flex-shrink` to each rating image as this will allow the image to shrink in order to fit the available space:

```
.rating {
    margin: 1rem auto;
    width: 60%;
    display: flex;
    flex-wrap: row nowrap;
}
```

```
.rating img {
    width: 20%;
    flex-shrink: 1;
}
```

8. Next, we will add `alt` attributes to each image element with an empty string. This means that the images will be treated as decorative:

```
<div class="rating">
    <img
        src="images/full-star.png"
        alt=""
    />
    <img
        src="images/full-star.png"
        alt=""
    />
    <img
        src="images/full-star.png"
        alt=""
    />
    <img
        src="images/half-star.png"
        alt=""
    />
    <img
        src="images/empty-star.png"
        alt=""
    />
</div>
```

The five image elements are required to describe some information that our users may find useful. At the moment, our rating element does not describe that information in a text-based format. A screen reader user would not get information about the product rating and that could hamper their choice.

9. By adding the rating information as the value of an `alt` tag for the first image, we provide the necessary information for non-visual users:

```
<div class="rating">
    <img
        src="images/full-star.png"
        alt="Rated 3 and a half out of 5
            stars"
    />
    <img
```

```
            src="images/full-star.png"
            alt=""
        />
        <img
            src="images/full-star.png"
            alt=""
        />
        <img
            src="images/half-star.png"
            alt=""
        />
        <img
            src="images/empty-star.png"
            alt=""
        />
    </div>
```

We've added `alt` attributes to each of the images, but only the first one describes the information portrayed by the set of images. This provides information for a user who cannot see the visual representation of the images and means that the ratings are accessible and useful.

In a browser, the final result will look like the screenshot that was provided at the beginning of this exercise.

The following screenshot shows the product page without images. The `alt` text shows that we can still get the information provided by the images when the images are not available. The rating is shown as **Rated 3 and a half out of 5 stars** so that a user who has a screen reader can make a decision about this product with the same information as a user who can see the images:

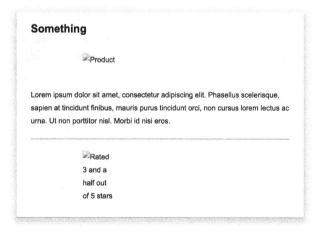

Figure 9.4: Product page with alt text replacing images

In the next section, we are going to look at what we can do to make HTML forms accessible to users.

Making forms accessible

We learned about HTML forms in *Chapter 4*. Making forms accessible is very important because forms are one of the key areas where users will interact with your site. This is where users will sign up, send feedback, or pay for goods.

Making forms accessible takes some thinking, and there are fundamental practices we should follow when we are creating forms for a web page. By following these practices, we will have gone a long way toward making accessible forms and web pages.

The techniques we will look at here are as follows:

- Labels and input fields
- Fieldsets

A common mistake that's found in forms across the web can be seen here:

```
<p>First name:</p>
<br />
<input type="text" id="first-name" />
```

The following screenshot shows the result of this markup:

First name:

Figure 9.5: Form markup

Visually, this markup may look fine. This may be how we've designed the input and label. However, the problem for a user of a screen reader is that the **First name:** text is not associated with the input. A screen reader would not read out the text when navigating to the form field.

We need to provide labels for our form inputs, and we need to make sure that the labels are correctly associated with those input fields.

To make the preceding example accessible, we can do one of two things:

- We can associate a text label with a form input field using a label element with a for attribute. The value of the for attribute must match an id attribute value on the appropriate form field. For example, the following code provides the **First name** label and it is associated with the input with the first-name ID:

```
<label for="first-name">First name</label>
<input type="text" id="first-name" />
```

- We can wrap the input with the label, again creating an association between the two. This would look like this:

```
<label>First name: <input type="text" /></label>
```

In both cases, we now have a form field associated with a label. When a screen reader user navigates to the input element with the first-name ID, the screen reader will read out the label **First name**. This explains what information the input field is expecting without the user relying on a visual association.

When forms get more complex with a large number of input fields, they can be difficult to navigate through and understand. We often use white space and visual groupings to break up a form, but that doesn't work if a user is browsing with a screen reader. To meaningfully break information up into understandable groupings, we can use the fieldset and legend elements.

A fieldset element wraps a set of form input fields. You can add a legend element nested in the fieldset element to provide a textual caption for the fieldset element. For a screen reader user, this grouping and text caption helps them understand what they are being asked to input and provides context.

As an example, the following code creates a form with two fieldsets. The first set of fields asks for the user's address and the second set asks the user to choose their favorite color. The use of fieldset helps all users understand that the two fieldsets are grouped separately:

```
<form>
    <fieldset>
        <legend>Provide your address:</legend>
        <label for="house">House</label>
        <input type="text" id="house" />
        <br />
        <label for="street">Street</label>
        <input type="text" id="street" />
        <br />
        <label for="zipcode">ZIP code</label>
        <input type="text" id="zipcode">
    </fieldset>
    <fieldset>
        <legend>Choose a favorite color:</legend>
        <input
            type="radio"
            value="red"
            id="red"
            name="color"
        >
        <label for="red">Red</label>
        <input
```

```
                    type="radio"
                    value="green"
                    id="green"
                    name="color"
            >
            <label for="green">Green</label>
            <input
                    type="radio"
                    value="blue"
                    id="blue"
                    name="color"
            >
            <label for="blue">Blue</label>
        </fieldset>
    </form>
```

The result of the preceding code (with default user agent styling) is shown in the following screenshot:

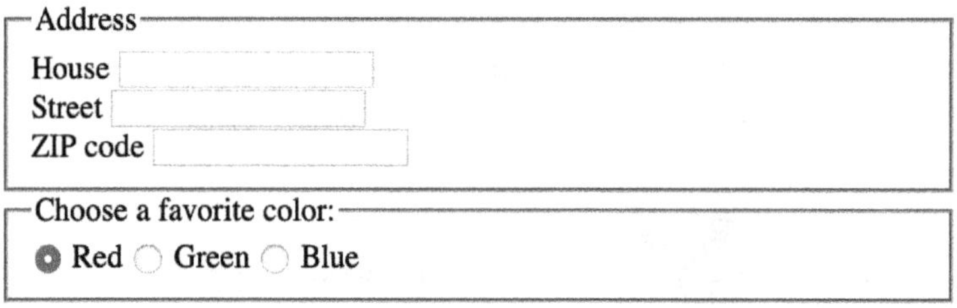

Figure 9.6: Form with fieldsets and legend

Visually, the two different fieldsets are obvious and thematically related to the input fields. For a screen reader, the second `fieldset` might be read as follows:

Choose a favorite color radio button Red, radio button Green, radio button Blue.

Having looked at some of the fundamental techniques we can use to make forms accessible, we will put these techniques into practice in the next exercise by creating an accessible sign up form.

Exercise 9.02: accessible sign up form

We are going to create an accessible HTML form in this exercise. The form will be a simple example of a sign up form and we will focus on making it accessible by providing the appropriate fields with labels and grouping them as fieldsets. By doing this, we will learn how to make an accessible form.

We can see the wireframe of the form in the following diagram. This is a sign up form; it is simple and functional, and we will particularly focus on making sure it is accessible by making use of the `label` and `fieldset` elements:

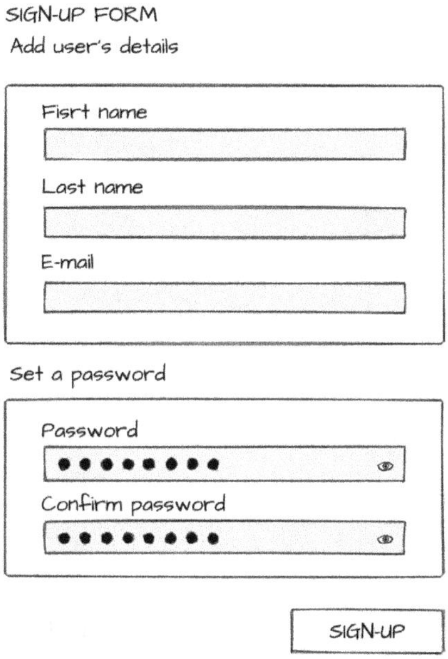

SIGN-UP FORM
Add user's details

Fisrt name

Last name

E-mail

Set a password

Password

● ● ● ● ● ● ● ●   ☺

Confirm password

● ● ● ● ● ● ● ●   ☺

SIGN-UP

Figure 9.7: Wireframe of the sign up form

As we can see from the wireframe, the form will have a heading (*Sign up Form*), two fieldsets, each with a legend (*Add user's details* and *Set a password*), a submit button with the *Sign up* label, and the five input fields with corresponding labels for *First name*, *Last name*, *email*, *Password*, and *Confirm Password*:

The steps are as follows:

1. First, we will create an empty web page in which to develop our accessible web form. We will create and save a file named `Exercise 9.02.html` and add the following code to set up a basic HTML page with a title and styles to remove white space around the body element:

```
<!DOCTYPE html>
<html lang="en">
    <head>
        <meta charset="utf-8">
        <title>
            Exercise 9.02: Accessible sign up form
        </title>
        <style>
        html {
```

```
            box-sizing: border-box;
        }

        *, *:before, *:after {
            box-sizing: inherit;
        }

        body {
            padding: 0;
            margin: 0;
            font-family: Arial, Helvetica, sans-serif;
            font-size: 16px;
            line-height: 1.7778;
        }
        </style>
    </head>
    <body>

    </body>
</html>
```

2. We want to create a form, so we'll add a `class` attribute with the `sign up` value, which we will use for styling. We will add a `h2` heading element to the form with the *Sign up Form* text content so that we know what the form is for:

```
<form class="signup">
    <h2>Sign up form</h2>

</form>
```

3. Next, we'll add our `input` fields to the form. The information we want to capture is the user's first and last name, an email address, and a password. We also need a button to submit the form. We will use the `label` element to provide the appropriate labeling for each field. To make sure each `label`/`input` pair is on a separate line, we are going to use the `br` element. We have connected each `label` to an `input` using the `for` attribute and a corresponding `id` on the `input` element:

```
<form class="signup">
    <h2>Sign up form</h2>
    <label for="first-name">
        First name:
    </label>
    <input type="text" id="first-name" />
    <br />
```

```
      <label for="last-name">Last name:</label>
      <input type="text" id="last-name" />
      <br />
      <label for="email">E-mail:</label>
      <input type="email" id="email" />
  <br />
  <label for="password">Password:</label>
      <input type="password" id="password" />
      <br />
      <label for="confirm-password">
          Confirm password:
      </label>
      <input
          type="password"
          id="confirm-password"
      />
      <br />
      <button
          type="submit"
          class="sign up-button"
      >
          Sign-up
      </button>
  </form>
```

The following screenshot shows our form at this stage:

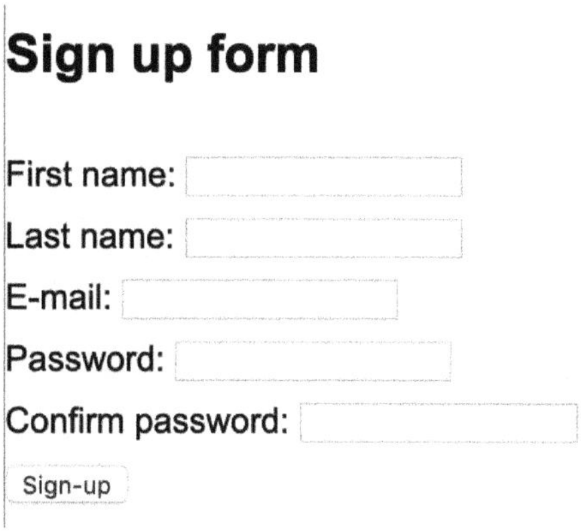

Sign up form

First name:
Last name:
E-mail:
Password:
Confirm password:
Sign-up

Figure 9.8: Unstyled form with inputs and labels

4.   We can improve the layout of the form to help make the experience better for visual users. Add the following CSS to the style element in order to add some white space around the form, style the **Sign up** `button` so that it's more visually interesting and so that it stands out more, and increase the size of the `input` fields so that they take up a whole line:

```css
.signup {
    margin: 1rem;
}

.signup-button {
    -webkit-appearance: none;
    appearance: none;
    padding: .5rem 1rem;
    background: #7e00f4;
    color: white;
    font-size: 1rem;
    border-radius: 3px;
    margin: 1rem 0;
    outline: none;
    float: right;
    cursor: pointer;
}

.signup-button:hover,
.signup-button:focus {
    background: #500b91;
}

.signup label {
    display: inline-block;
    width: 150px;
}

.signup input {
    width: 100%;
    height: 2rem;
    margin-bottom: 1rem;
}
```

The following screenshot shows the effect of styling the form with the preceding CSS code. It makes the form more visually appealing and easier to work with:

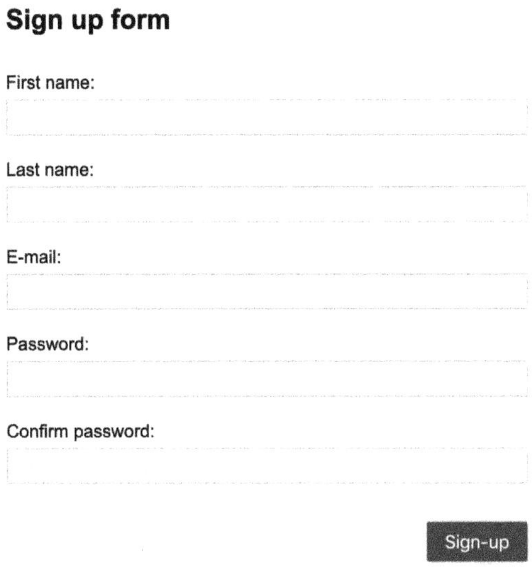

Sign up form

First name:

Last name:

E-mail:

Password:

Confirm password:

Sign-up

Figure 9.9: Styled sign up form

5. All our users will benefit from partitioning the form thematically. It will make it easier to understand for visual and non-visual browsers alike. We will do this by separating the password and user data portions with the `fieldset` element. We will give each part a `legend` element to describe the information being requested.

The first `fieldset` tag is for user details:

```
<fieldset>
    <legend>Add user's details:</legend>
    <label for="first-name">
        First name:
    </label>
    <input type="text" id="first-name" />
    <br />
    <label for="last-name">
        Last name:
    </label>
    <input type="text" id="last-name" />
    <br />
    <label for="email">E-mail:</label>
    <input type="email" id="email" />
</fieldset>
```

The second `fieldset` tag is for setting a password:

```
<fieldset>
    <legend>Set a password:</legend>
    <label for="password">Password:</label>
    <input type="password" id="password" />
    <br />
    <label for="confirm-password">
        Confirm password:
    </label>
    <input
        type="password"
        id="confirm-password"
    />
</fieldset>
```

6. To finish, we add some styling to the `fieldset` element by adding the following declaration to the style element:

```
.sign up fieldset {
    margin: 1rem 0;
    background: #f5ffff;
    border: 2px solid #52a752;
}
```

The end result will be similar to what is shown in the following figure. The screenshot shows the final output – an accessible sign up form:

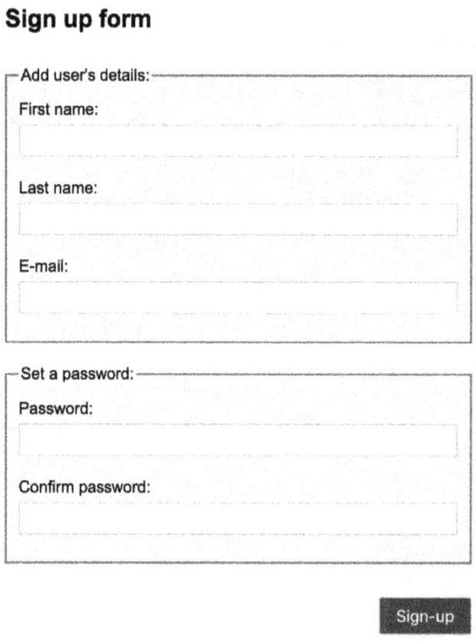

Figure 9.10: The accessible sign up form we have created

This form has been labeled and the labels are paired appropriately, to form fields. The form fields are grouped to help a user make sense of them. All users will benefit from these structural grouping and labeling techniques and they will allow screen reader users to sign up successfully.

In this exercise, we have created an accessible form and learned about grouping parts of a form and associating labels with inputs to make sure users can access the form using a screen reader.

In the next section, we will look at keyboard accessibility. The keyboard provides input for many users who cannot use a mouse. As developers, we cannot take mouse usage for granted. Many users find the mouse difficult or impossible to use and many use the keyboard out of necessity or as a preference.

Understanding keyboard accessibility

For some of our web page's users, a mouse may not be of much use. It requires the ability to follow a visual pointer onscreen, a certain amount of sensitivity of touch, and fine motor skills.

Whenever we test our web pages, we should check that we can use the keyboard to get to all the content and that we can interact successfully without using a mouse.

We can navigate through a web page using the following keys on the keyboard:

- *Tab*
- *Shift + Tab*
- *Enter*

To navigate a web page, we use the *Tab* key to cycle through elements on the web page. We can see this if we open the Exercise 9.02.html file and use the *Tab* key. *Shift + Tab* will cycle in the reverse order. We can use the two in combination to move back and forth through the elements of the web page.

In the following screenshot, you can see the default focus ring on the **First name** input field:

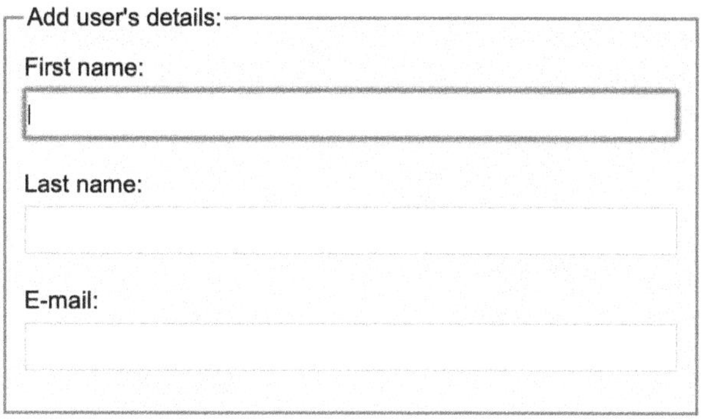

Figure 9.11: Form with a focus on the First name field

In the following screenshot, we see that the focus ring has moved to the **Last name** input. This is because we pressed the *Tab* key:

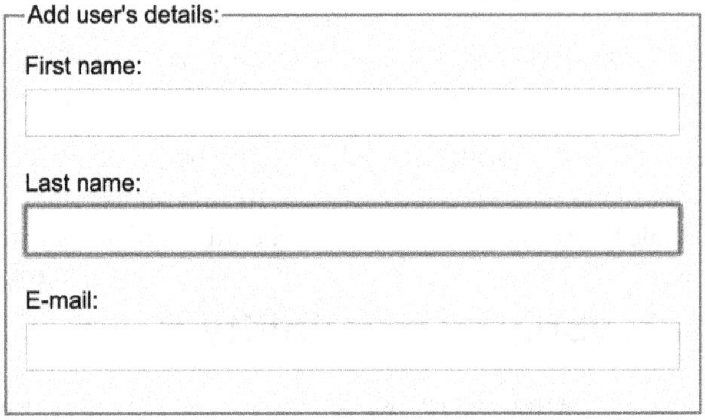

Figure 9.12: Form with a focus on the Last name field

To submit a form or interact with a link or button, we can use the *Enter* key. In the case of buttons, we can also use the spacebar to use a button.

To toggle radio and checkbox input fields, we can use the spacebar.

Checking that styling or JavaScript has not made the tab order confusing is very important for keyboard accessibility.

We should also make sure that our interactive elements, such as buttons and input fields, have some differentiation for when they have focus. We can do this with the :focus pseudo-class in CSS.

Making sure you can consistently interact with the elements on your web page using a keyboard helps you know with confidence that a user will not find barriers when using your web page.

In this section, we have learned about the importance of making a web page accessible with the keyboard and how to use keyboard controls to check that we can access interactive elements on our web page.

Next, we will look at how we can use a media query provided through CSS to handle animation responsibly. Motion can be a great addition to a web page but can also cause distraction and, in some cases, distress. By following user preferences, we can make sure motion does not prevent a user from enjoying our web page.

Making motion accessible

CSS animations and transitions have become a massive part of creating rich and visually interesting web pages. We learned about animating our web pages in *Chapter 5*. Motion can add a lot to a web page and can help us highlight content or make content changes more obvious.

With power comes responsibility, and in the case of motion, we should consider whether repeated transitions and animations may cause a user distraction and, in some cases, irritation or make for a difficult user experience.

In particular, repetitive motion, parallax effects, and flicker effects can be really bad for people with vestibular disorders, where such motion can cause severe discomfort and even nausea.

As browsers get better integration with operating systems, they can make use of more fine-grained accessibility configurations. In the case of motion, a media query has been added to most modern browsers, which we can use to check whether the user prefers to have reduced levels of motion. We can combine this, responsively, with our CSS transitions and animations to give a user the appropriate experience.

The media query is `prefers-reduced-motion` and it has two possible values: `reduce` or `no-preference`.

For example, the following CSS would apply an animation only if the user has no preference regarding the `prefers-reduced-motion` media query:

```
<div class="animation">animated box</div>
<style>
    .animation {
        position: absolute;
        top: 150px;
        left: 150px;
    }

    @media (prefers-reduced-motion: no-preference) {
        .animation {
            animation: moveAround 1s 0.3s linear infinite
                       both;
        }
    }

    @keyframes moveAround {
        from {
            transform: translate(-50px, -50px);
        }
        to {
            transform: translate(50px, 100px);
        }
    }
</style>
```

We can test this media query by configuring the accessibility setting in our operating system. To do so, do the following:

- On Mac, go to **System Preferences** > **Accessibility** > **Display** and toggle the **Reduce motion** checkbox
- On Windows, this preference can be controlled via **Settings** > **Ease of Access** > **Display** > **Show Animations**.

The following screenshot shows the **Accessibility** display preferences on Mac with the setting for **Reduce motion** selected:

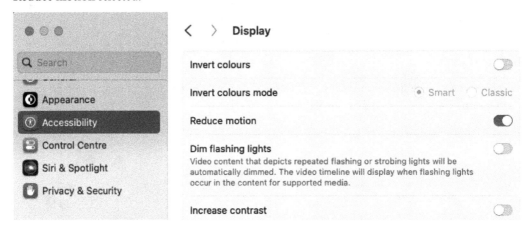

Figure 9.13: Mac display preferences with the Reduce motion checkbox

In the next section, we will look at how we can use tools to help understand and audit a web page for accessibility problems.

Accessibility tools

There are a lot of built-in, third-party tools available that can help with different aspects of accessibility.

A great number of tools have been created that can help us with the accessibility of our web pages. Some are good for diagnosing issues and auditing pages for structural issues, some help us check tab order and keyboard accessibility, and some help us with design decisions around color contrast and text legibility.

In this section, we are going to look at the Axe accessibility checker, which can be used to audit a web page or site to highlight issues in our semantics and HTML structure that may cause accessibility issues.

The Axe® tool

Axe® is an accessibility tool developed by Deque. It is a popular accessibility testing resource that can be used to flag issues on our web pages. There are several versions of this tool, but we are going to look at the free Chrome extension, "*axe DevTools - Web Accessibility Testing*".

To install the extension, we can go to the Chrome web store at `https://packt.link/t8KFq`.

If you are running the tool on a local file (`file://`), you may have problems with Axe throwing an error. The solution is to do the following:

1. In your browser, navigate to `chrome://extensions`.
2. Find the Axe extension (with the **axe - Web Accessibility Testing** heading).
3. Click the **Details** button.
4. Locate the **Allow access to file URLs** switch and enable it.

This will allow you to run the Axe extension on local web pages.

The tool is added as an **axe** tab in the Chrome web developer tools. Open the web developer tools and select the **axe** tab. You will see the panel shown in the following screenshot:

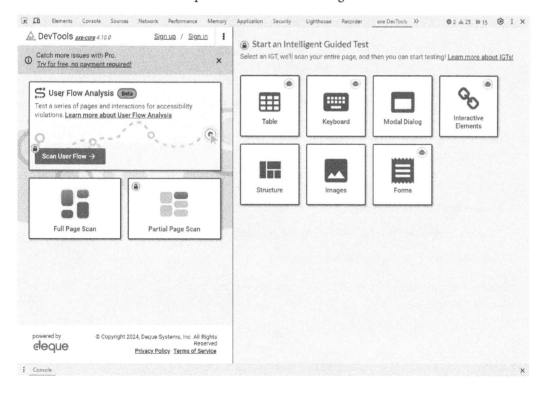

Figure 9.14: Axe accessibility checker Chrome extension

To run the accessibility checker, click the **Full Page Scan** button in the left-hand panel. This will check the site we are currently on and report any issues it has detected from analyzing the markup.

The following screenshot shows the reported results from running Axe on a web page.

Please note that since the creation of this content, the Axe user interface has undergone significant changes. As a result, the information may now appear in a different layout or location. In the left-hand panel, we can see a summary of all the issues that were found. These can be anything from semantic HTML issues, through incorrect levels of nested headings, to color contrast issues. The right-hand panel provides a more detailed description of an issue, including the location of the issue and what can be done to fix the issue:

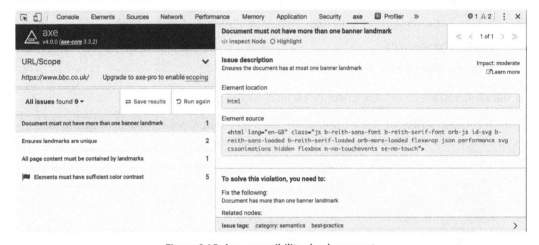

Figure 9.15: Axe accessibility checker report

Taking the preceding screenshot as an example, we can see that four different issues have been reported for the analyzed page. The issues are as follows:

- **Document must not have more than one banner landmark**
- **Ensures landmarks are unique**
- **All page content must be contained by landmarks**
- **Elements must have sufficient color contrast**

Several of these refer to landmarks, which we learned about in *Chapter 1*. The results are flagged for the following reasons:

- An HTML page should only have one banner landmark (meaning the HTML header element or the **WAI-ARIA** `role="banner"` attribute). This is because it makes the web page easier for non-visual users to navigate and find their way around.

- A landmark should be unique, which means it should have a unique combination of role and label. This helps to distinguish between landmarks. Again, this improves navigation around the web page.

- All the content of a page should be contained within a landmark (such as `header`, `footer`, `nav`, or `role="banner"`), and again, this is important for web page navigation.

- Strong color contrast helps users to distinguish text from its background and helps all visual users, especially those with visual impairments. Axe can detect low contrast levels between a foreground (text color) and background color and will flag such issues.

These are just a few of the many topics that the Axe tool can and will flag for you.

In this section, we've looked at the Axe tool and some of the results it flags. Next, we will try it out on a web page of our own.

Exercise 9.03: Using Axe

In this exercise, we will run the Axe tool on a web page and analyze the findings in the report it creates.

The steps are as follows:

1. First, create and save a file named `Exercise 9.03.html`. In `Exercise 9.03.html`, copy and paste the following code to set up an HTML page. The page is the bare bones of quite a simple layout and contains some landmarks (such as `nav`, `header`, and `footer`). The page has the following markup:

```
<!DOCTYPE html>
<html lang="en">
    <head>
        <meta charset="utf-8">
        <title>Exercise 9.03: Using Axe</title>
        <style>
        html {
            box-sizing: border-box;
        }

        *, *:before, *:after {
            box-sizing: inherit;
        }

        body {
            padding: 0;
            margin: 0;
            font-family: Arial, Helvetica, sans-serif;
            font-size: 16px;
```

```css
        line-height: 1.7778;
}

.container {
    height: 100vh;
    display: flex;
    align-items: center;
    justify-content: flex-start;
    flex-flow: column nowrap;
}

.header, .footer, [role="banner"] {
    background: black;
    color: white;
    width: 100vw;
    padding: 0 1rem;
}

h1, h2, h3, h4, h5, h6 {
    padding: 0;
    margin: 0;
}

[role="banner"] {
    background: yellow;
}

.content {
    flex: auto 1;
}

.nav ul {
    margin: 0;
    padding: 1rem 0;
}

.nav li {
    display: inline-block;
    padding: 0 1rem;
    list-style: none;
}
```

```
    .nav a,
    .nav a:link,
    .nav a:visited,
    .nav a:hover,
    .nav a:active,
    .nav a:focus {
        color: white;
        text-decoration: none;
    }

    .nav a:focus,
    .nav a:hover {
        text-decoration: underline;
    }
    </style>
</head>
<body>
    <div class="container">
        <header class="header">
            <h1>Site Heading</h1>
        </header>
        <div role="banner">
            <nav class="nav">
                <ul>
                    <li>Nav 1</li>
                    <li><a href="#">Nav 2</a></li>
                    <li><a href="#">Nav 3</a></li>
                </ul>
            </nav>
        </div>
        <div class="content">
            <h2>Content Heading</h2>
            Lorem ipsum...
        </div>
        <footer class="footer">
            <h3>Site Links:</h3>
            <nav class="nav">
                <ul>
                    <li>Nav 1</li>
                    <li><a href="#">Nav 2</a></li>
                    <li><a href="#">Nav 3</a></li>
                </ul>
            </nav>
```

```
          </footer>
        </div>
      </body>
    </html>
```

2. We will open this page in the browser where we have the Axe extension. The following screenshot shows this page:

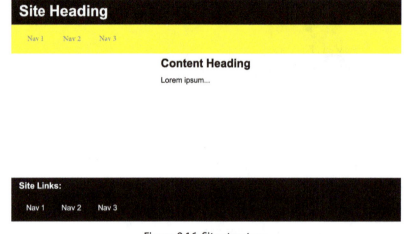

Figure 9.16: Site structure

3. Next, open the developer tools and select the **axe** DevTools tab. Then, run the Axe analysis tool by clicking the button labeled **Scan ALL of my page**.

4. After the scan, you will see an overview with three serious issues flagged. These issues will all be of the type **Elements must have sufficient color contrast**, with a description of the issue and a pointer to the element location.

5. If we set the **Best Practices** toggle to **ON**, we will see a further five moderate issues, resulting in eight issues in total, with the following types of issues all represented:

    I.   **Elements must have sufficient color contrast**

    II.   **Document must not have more than one banner landmark**

    III.   **Document must have one main landmark**

    IV.   **Ensures landmarks are unique**

    V.   **All page content must be contained by landmarks**

    We will look at each issue one by one. The details of the first result are selected by default.

As shown in the following screenshot, we have several elements with insufficient color contrast. From the issue description, we can see that this is because the top nav tag has a yellow background and white text. The text is barely visible:

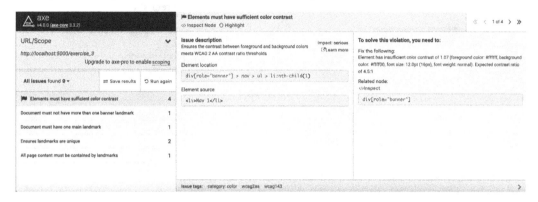

Figure 9.17: Details of the Elements must have sufficient color contrast issue

6.   Select the next result (**Document must not have more than one banner landmark**). This issue is caused by having both a header element and a div element with role="banner". Both of these landmarks have the same purpose and role in an HTML page. We can see the suggested solutions in the following screenshot:

Figure 9.18: Details of the Document must not have more than one banner landmark issue

7.   Next, we select the third result (**Document must have one main landmark**). This issue is caused by having no main element or element with `role="main"` included on the page. We have a `div` tag with `class="content"`, but we could benefit our users by making this a main element. Again, we see the suggested solution in the following screenshot:

Figure 9.19: Details of the Document must have one main landmark issue

8.   Next, we select the fourth result (**Ensure landmarks are unique**). This issue is similar to the second one and is caused by the same problem (that we have a header and an element with `role="banner"`). This means that the two landmarks are not unique. We can see the solution in the following screenshot:

Figure 9.20: Details of the Ensure landmarks are unique issue

9. Next, we select the final result (**All page content must be contained by landmarks**). The cause of this issue is the content being in a `div` element that is not a landmark (the `div` with class content). Again, we can fix this by changing this `div` to a main element. We can see the solution in the following screenshot:

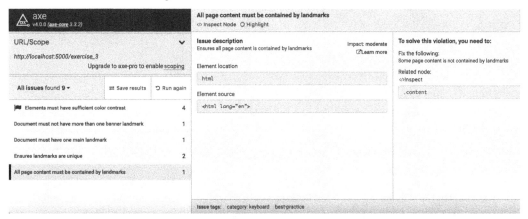

Figure 9.21: Details of the All page content must be contained by landmarks issue

By running the Axe tool, we have found some issues early in the development of a web page, when they can be easily solved. The Axe tool has helped us discover these issues quickly and has suggested solutions.

In this exercise, we have seen how the Axe web accessibility checker can help us identify structural, semantic, and visual accessibility issues in our web pages.

To put into practice all the skills we have learned in this chapter, we will finish with an activity in which we will use the Axe tool to diagnose some issues with a web page. We will then fix the issues that have been flagged for us.

Activity 9.01: Making a page accessible

You have been asked by a client to look at the accessibility of their product feedback page. The client has received several complaints from users trying to send feedback. The client has provided the source code for the page and wants you to make changes to improve the accessibility of the page.

The steps are as follows:

1. First, make a copy of the `activity_1_inaccessible.html` file and rename it `Activity 9.01.html`. The code for `activity_1_inaccessible.html` is at `https://packt.link/1RJ7Z`.

2. Run the Axe accessibility checker tool on this page.

3. In the Axe tool's summary, you should see five types of issues flagged by the Axe tool.

4. Fix each of the issues with the techniques you have learned about in this chapter and the hints that are given by the Axe tool.

5. Then, run the Axe tool again to check that the issues have been fixed.

Summary

In this chapter, we learned about accessibility. We looked at some of the simple techniques we can employ to make our web pages accessible. We also learned about the right ways to use `alt` text and text descriptions to make the images we embed in a web page accessible to users who can't see the images. We then learned about making forms accessible and about controlling a web page via the keyboard.

Most importantly, we raised awareness about a topic that affects a lot of people and an area that all web developers should champion.

In the next chapter, we will look at how build tools can help us with developing complex modern websites and how we can use preprocessors to expand the capabilities of what we can do in web development.

Part 4:
Advanced Concepts

In this part, we'll go the extra mile in web development by exploring advanced techniques to elevate your skills. We'll begin by examining how to enhance your application with SEO strategies, preprocessors, and efficient tooling. These concepts will equip you with powerful resources to take your projects to the next level. Next, we'll focus on strategies for maintaining CSS – a critical skill as projects grow in complexity. Finally, we'll investigate the latest advancements in web development and discover how cutting-edge technologies can further enhance our applications.

This section contains the following chapters:

- *Chapter 10, SEO Essentials for Web Developers*
- *Chapter 11, Preprocessors and Tooling for Efficient Development*
- *Chapter 12, Strategies for Maintaining CSS Code*
- *Chapter 13, The Future of HTML and CSS – Advancements and Trends*

10

SEO Essentials for Web Developers

Understanding **Search Engine Optimization (SEO)** is crucial in today's digital landscape, especially for web developers who aim to create impactful and high-performing websites. This chapter will delve into the essentials of SEO from a web developer's perspective. We will explore how **Hypertext Markup Language (HTML)** and **Cascading Style Sheets (CSS)**can be optimized for better search engine visibility and user experience.

By understanding how search engines crawl, index, and rank web pages, developers can make informed decisions that optimize site structure, content, and performance. Let's explore the relationship between web development and SEO, emphasizing how a well-optimized site can drive organic traffic, improve user experience, and ultimately contribute to the success of any online presence.

In this chapter, we'll cover the following topics:

- Understanding SEO and its importance
- Exploring Core Web Vitals
- Minimizing render-blocking resources
- Optimizing images for SEO
- Methods to reduce **Document Object Model (DOM)** size for improved performance

By mastering these topics, you'll be equipped to significantly improve your website's SEO, boosting its visibility, performance, and user experience.

Technical requirements

The code files for this chapter can be found at `https://packt.link/TEToe`.

Understanding SEO and its importance

SEO is the practice of enhancing a website to increase its visibility in **Search Engine Results Pages** (**SERPs**). When users search for keywords related to specific content, a well-optimized site will appear higher in the results, driving more organic (non-paid) traffic to the site. SEO involves a range of techniques and best practices, including keyword research, on-page optimization, technical SEO, and link building.

As a web developer, understanding SEO and implementing its best practices is essential for enhancing your application's visibility on search engines. Effective SEO ensures that your site ranks well, making it easier for users to find. As the saying goes, "The best place to hide a secret is on the second page of Google search results." This highlights the importance of securing a top spot in search rankings.

Key SEO concepts for web developers

To fully leverage SEO strategies, it's essential to understand the various terms and key concepts involved, such as the following:

- **Keyword research**: Identifying the terms and phrases that potential visitors use to find relevant content is crucial.

- **On-page optimization**: Optimizing individual pages to rank higher and earn more relevant traffic is very important. This includes title tags, meta descriptions, header tags, and keyword placement.

- **Technical SEO**: Improving the technical aspects of a website, such as site speed, mobile-friendliness, indexing, crawlability, and security, is necessary.

- **Link building**: It's important to acquire hyperlinks from other websites to your own. High-quality backlinks can significantly improve a site's authority and ranking.

- **Content creation**: Producing high-quality, valuable content that answers users' queries and encourages engagement and sharing is a major part of SEO optimization.

By mastering SEO, web developers can build websites that not only function well but also attract and retain a broad audience, ultimately leading to the success and growth of their projects.

Benefits of developing with SEO practices in mind

By integrating SEO practices into web development, you not only improve search engine rankings but also create a more user-friendly, efficient, and successful website. This holistic approach ensures that your site meets both search engine criteria and user expectations, driving sustained growth and success. Here are some benefits of SEO for developers:

- **Visibility**: Higher rankings in search results lead to greater visibility. Websites that appear on the first page of search results get more clicks and views than those on subsequent pages.

- **Traffic**: More visibility means more traffic. Websites that are optimized for search engines attract more organic traffic, which can lead to increased conversions and revenue.

- **Site structure**: SEO encourages a well-structured website, making it easier for users to navigate. This includes clear sitemaps, logical URL structures, and organized content.

- **Load speed**: Part of SEO involves optimizing site speed. Faster-loading pages enhance user experience and reduce bounce rates.

- **Mobile performance**: With a significant portion of web traffic originating from mobile devices, search engines often use **mobile-first indexing**, which leads developers to embrace SEO strategies to enhance both user experience and performance. You can learn more about how Google indexes using mobile-first at `https://developers.google.com/search/docs/crawling-indexing/mobile/mobile-sites-mobile-first-indexing`.

- **Cost-effective marketing**: SEO is one of the most cost-effective marketing strategies. Unlike paid advertising, organic traffic is free, and a well-optimized site can continue to attract visitors over time without incurring ongoing costs.

- **Higher conversion rates**: SEO-targeted traffic is often more qualified, meaning visitors are more likely to convert into customers or leads.

- **Industry standards**: SEO is a standard practice in web development and digital marketing. Staying competitive requires understanding and implementing SEO strategies to match or surpass competitors.

- **Algorithm updates:** Search engines frequently update their algorithms. Web developers need to stay informed about these changes to maintain or improve their site's rankings.

Given all these benefits, it's essential to understand the roles played by the elements of web applications. Let's begin by exploring how HTML and CSS impact SEO.

The impact of HTML and CSS on SEO

HTML is the backbone of web content and plays a crucial role in SEO. Here's how:

- **Semantic HTML tags**: Semantic HTML tags (such as `<header>`, `<footer>`, `<article>`, and `<section>`) help search engines understand the structure and content of your web pages. Proper use of these tags makes it easier for search engines to index and rank your content appropriately. Here's an example of a semantic HTML structure:

```
<!DOCTYPE html>
<html lang="en">
<head>
    <meta charset="UTF-8">
    <meta
        name="viewport"
```

```
            content="width=device-width,
                initial-scale=1.0"
        >
    <title>My Web Page</title>
</head>
<body>
    <header>
        <h1>Welcome to My Website</h1>
        <nav>
            <ul>
                <li><a href="#home">Home</a></li>
                <li><a href="#about">About</a></li>
                <li><a href="#contact">
                    Contact
                </a></li>
            </ul>
        </nav>
    </header>
    <main>
        <article>
            <section>
                <h2>About Us</h2>
                <p>
                    This section provides information
                    about our company.
                </p>
            </section>
            <section>
                <h2>Our Services</h2>
                <p>
                    Details about the services we
                    offer.
                </p>
            </section>
        </article>
    </main>
    <footer>
        <p>
            &copy; 2024 My Company. All rights
            reserved.
        </p>
    </footer>
</body>
</html>
```

- **Title tags**: A well-crafted `<title>` tag with relevant keywords can significantly improve your page's ranking and **Click-Through Rate** (**CTR**) in search results. Here is an example of `<title>` tags:

```
<!DOCTYPE html>
<html lang="en">
<head>
    <meta charset="UTF-8">
    <meta
    name="viewport"
        content="width=device-width,
                initial-scale=1.0"
    >
    <title>Top SEO Tips for Web Developers</title>
</head>
<body>
    <!-- Page content -->
</body>
</html>
```

- **Meta descriptions**: Meta descriptions (using the `<meta>` tag) provide a summary of the page content. While this is not a direct ranking factor, a compelling meta description can improve CTR, indirectly influencing rankings. Here's an example of this meta tag:

```
<!DOCTYPE html>
<html lang="en">
<head>
    <meta charset="UTF-8">
    <meta
        name="viewport"
        content="width=device-width,
                initial-scale=1.0"
    >
    <title>Top SEO Tips for Web Developers</title>
    <meta
        name="description"
        content="Discover the best SEO practices for
                web developers to enhance site
                visibility and performance in search
                engines."
    >
</head>
<body>
```

```
        <!-- Page content -->
</body>
</html>
```

- **Header tags**: Header tags (<h1>, <h2>, <h3>, etc.) structure your content and indicate its hierarchy. Proper use of header tags improves readability and helps search engines understand the main topics of your content. An example of header tags is as follows:

```html
<!DOCTYPE html>
<html lang="en">
<head>
    <meta charset="UTF-8">
    <meta
        name="viewport"
        content="width=device-width,
                initial-scale=1.0"
    >
    <title>Top SEO Tips for Web Developers</title>
</head>
<body>
    <header>
        <h1>Top SEO Tips for Web Developers</h1>
    </header>
    <main>
        <section>
            <h2>Understanding SEO Basics</h2>
            <p>
                Learn about the fundamental principles
                of SEO.
            </p>
        </section>
        <section>
            <h3>On-Page SEO Techniques</h3>
            <p>
                Explore various on-page optimization
                strategies.
            </p>
        </section>
    </main>
</body>
</html>
```

- **Internal linking**: Links between pages on your site help establish site architecture and hierarchy. Effective internal linking distributes link equity (ranking power) throughout your site, aiding in the indexing and ranking of pages. Here's an example of internal linking:

```html
<!DOCTYPE html>
<html lang="en">
<head>
    <meta charset="UTF-8">
    <meta
        name="viewport"
        content="width=device-width,
                 initial-scale=1.0"
    >
    <title>Top SEO Tips for Web Developers</title>
</head>
<body>
    <header>
        <h1>Top SEO Tips for Web Developers</h1>
    </header>
    <main>
        <section>
            <h2>Understanding SEO Basics</h2>
            <p>
                Learn about the fundamental principles
                of SEO.
            </p>
            <a href="seo-basics.html">
                Read more about SEO Basics
            </a>
        </section>
        <section>
            <h2>Advanced SEO Techniques</h2>
            <p>
                Delve into advanced strategies for
                optimizing your site.
            </p>
            <a href="advanced-seo.html">
                Explore Advanced SEO Techniques
            </a>
        </section>
    </main>
</body>
</html>
```

- **URL structure:** Clean, descriptive URLs make it easier for users and search engines to understand page content. Including keywords in URLs can improve CTRs and rankings. Here's an example:

```
<!DOCTYPE html>
<html lang="en">
<head>
    <meta charset="UTF-8">
    <meta
        name="viewport"
        content="width=device-width,
                initial-scale=1.0">
    <title>Top SEO Tips for Web Developers</title>
</head>
<body>
    <header>
        <h1>Top SEO Tips for Web Developers</h1>
    </header>
    <main>
        <section>
            <h2>Understanding SEO Basics</h2>
            <p>
                Learn about the fundamental principles
                of SEO.
            </p>
            <a
                href="https://www.example.com/
                seo-basics"
            >
                Read more about SEO Basics
            </a>
        </section>
    </main>
</body>
</html>
```

- **Canonical tags:** Canonical tags (`<link rel="canonical" href="URL">`) indicate the preferred version of a page. They help prevent duplicate content issues, ensuring search engines index the correct version. Here's an example of canonical tags:

```
<!DOCTYPE html>
<html lang="en">
<head>
    <meta charset="UTF-8">
    <meta
```

```
        name="viewport"
        content="width=device-width,
              initial-scale=1.0"
    >
    <title>Top SEO Tips for Web Developers</title>
    <link
        rel="canonical"
        href=https://www.example.com/top-seo-tips
    >
</head>
<body>
    <header>
        <h1>Top SEO Tips for Web Developers</h1>
    </header>
    <main>
        <section>
            <h2>Understanding SEO Basics</h2>
            <p>
                Learn about the fundamental principles
                of SEO.
            </p>
        </section>
    </main>
</body>
</html>
```

As we've discussed, HTML code serves as a fixed representation of the application's dynamics for search engines and is one of the most impactful elements in SEO. Next, we'll explore how CSS can also influence SEO and how to optimize it for better results.

CSS and SEO

CSS primarily focuses on the presentation of web content. While it does not directly affect SEO, it has several indirect impacts:

- **Page load speed**: The way in which CSS is implemented can influence page load times. Faster-loading pages are favored by search engines and provide a better user experience, reducing bounce rates.

- **Mobile friendliness**: CSS allows for responsive design, ensuring websites are usable on various devices. Mobile-friendly websites rank better in mobile search results, as search engines prioritize mobile usability.

- **Content visibility**: CSS can be used to hide or display content. Hiding content with CSS (such as `display:none`) can be viewed suspiciously by search engines if it appears manipulative, potentially impacting rankings.

- **Font and readability**: CSS controls typography and layout. Better readability and a clean layout enhance user engagement, indirectly benefiting SEO through improved user behavior metrics.

- **Use of external CSS files**: External CSS files help keep HTML clean and reduce page size. Clean HTML is easier for search engines to crawl, and smaller page sizes improve load times.

- **Minification and compression**: Minifying and compressing CSS files reduces their size. This optimization improves page speed, which is a known ranking factor. *Figure 10.1* displays a CSS file in its default formatting alongside its minified version.

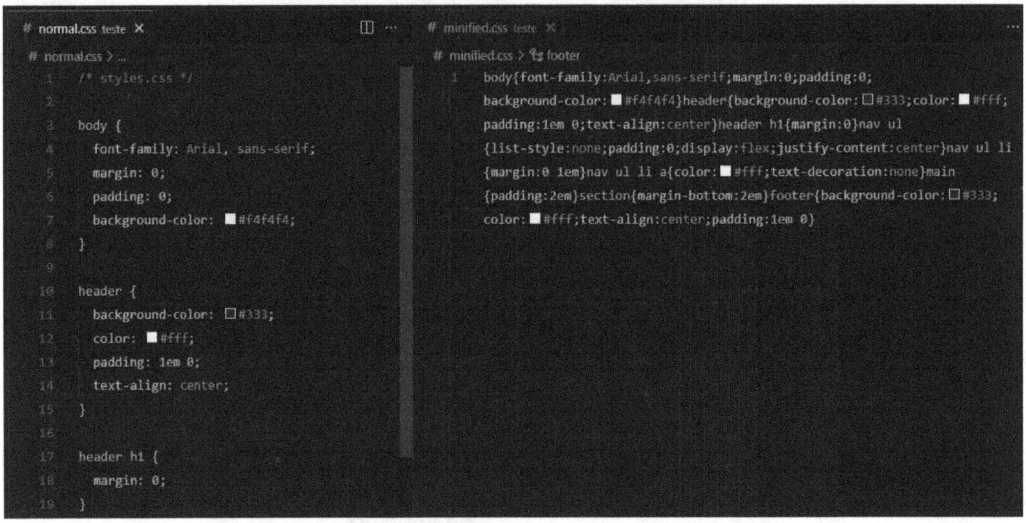

Figure 10.1 – Representation of a normal CSS file on the left, and minified CSS on the right

By effectively leveraging HTML and CSS, web developers can create websites that are not only visually appealing but also optimized for search engines, leading to better rankings, more traffic, and improved user engagement.

Let's summarize what we have learned in this section. Basically, to optimize your website for SEO, it's essential to focus on three key areas: **HTML structure**, **CSS performance**, and **accessibility**:

- First, ensure that your HTML structure is optimized by using semantic tags and maintaining a logical hierarchy. Additionally, create unique and descriptive title tags, as well as meta descriptions, for each page.

- Second, improve CSS performance by minifying CSS files and using external stylesheets, while also implementing responsive design principles to ensure that your site is mobile-friendly.

- Lastly, enhance accessibility by using alt attributes for images, ensuring all content is accessible to users with disabilities, and avoiding the use of CSS to hide content unless it is necessary for the user experience.

By adhering to these best practices, you can create a well-optimized, user-friendly website that performs well in search engine rankings.

Exercise 10.1 – adapting the HTML for SEO

In this exercise, we'll apply the concepts learned. Improve the given HTML and CSS code to follow SEO best practices, enhancing the website's visibility and performance:

1. To start the exercise, create a folder named `chapter10-SEO` within your `documents` directory. Open this folder using your preferred code editor, such as VS Code. Inside `chapter10-SEO`, create a new file called `index.html`. Paste the initial HTML code provided here into `index.html`:

```html
<!DOCTYPE html>
<html>
<head>
    <title>My Website</title>
    <style>
    body {font-family: Arial, sans-serif;}
        .header { font-size: 2em; }
        .content { margin: 20px; }
        .footer { font-size: 0.8em; }
    </style>
</head>
<body>
    <div class="header">Welcome to My Website</div>
    <div class="content">
        <div>
            <h2>About Us</h2>
            <p>We are a company that values...</p>
        </div>
        <div>
            <h2>Services</h2>
            <p>
                We offer a variety of services
                including...
            </p>
        </div>
    </div>
    <div class="footer">
```

```
                        Contact us at info@mywebsite.com
            </div>
        </body>
        </html>
```

This initial code serves as the starting point for implementing SEO best practices. Now, let's improve it step by step and check the resulting code at the end of the exercise.

If you require assistance with writing the code, you can access the source code by downloading it from our *GitHub* repository. You can find the link to the repository at `https://packt.link/qnp5G`.

Now, use semantic HTML elements in your file. Replace generic `<div>` tags with semantic elements such as `<header>` for `class="header"`, `<main>` for `class="content"`, `<section>` for elements used to separate content by theme, and `<footer>` for `class="footer"`.

2. After that, let us improve the `<title>` tag by making it more descriptive and including relevant keywords.

3. Remember to add `<meta>` tags for description and viewport.

4. Optimize headings by ensuring that they are structured correctly with `<h1>` for the main title and `<h2>` for subsections.

5. Finally, remove any inline styles that are not critical and place them in an external CSS file to reduce the DOM size and organize the code.

The result should match the following code:

```
<!DOCTYPE html>
<html lang="en">
<head>
    <meta charset="UTF-8">
    <meta
        name="viewport"
        content="width=device-width,
                initial-scale=1.0"
    >
    <meta
        name="description"
        content="My Website offers a variety of services
                including web development, SEO
                optimization, and digital marketing."
    >
    <title>My Website - Quality Services for You</title>
    <link rel="stylesheet" href="styles.css">
</head>
```

```
<body>
    <header class="header">Welcome to My Website</header>
    <main class="content">
        <section>
            <h1>About Us</h1>
            <p>
                We are a company that values quality and
                customer satisfaction. Our mission is to
                provide the best services to our clients...
            </p>
        </section>
        <section>
            <h2>Services</h2>
            <p>
                We offer a variety of services including
                web development, SEO optimization, and
                digital marketing...
            </p>
        </section>
    </main>
    <footer class="footer">
        Contact us at info@mywebsite.com
    </footer>
</body>
</html>
```

The external `styles.css` file should look like this:

```
body {
    font-family: Arial, sans-serif;
}

.header {
    font-size: 2em;
}

.content {
    margin: 20px;
}

.footer {
    font-size: 0.8em;
}
```

By following these steps, you have successfully optimized the HTML and CSS for enhanced SEO performance, thereby improving the visibility and user-friendliness of your website. Congratulations!

Exploring Core Web Vitals

Core Web Vitals comprises a set of metrics that assess real-world user experience in terms of loading performance, interactivity, and visual stability of a webpage.

Focusing on Core Web Vitals is important for both **improving user experience** and achieving **better search rankings**. Prioritizing these metrics helps ensure that webpages load quickly, respond promptly to user interactions, and maintain visual stability, all of which contribute to a positive user experience and align with SEO best practices. Let's delve deeper into these metrics.

First Contentful Paint

First Contentful Paint (FCP) is a crucial web performance metric that measures the time it takes for the browser to render the first piece of content (text, image, or other DOM elements) on the screen from the initial navigation request. It marks the point in the loading process when users can start seeing content on the webpage. *Figure 10.2* illustrates the FCP metric. Values of 1.8 seconds or less are considered good, while values exceeding 1.8 seconds can negatively impact the overall application score. The metric is as follows:

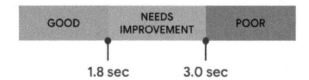

Figure 10.2 – Optimal FCP values are 1.8 seconds or below, while values
exceeding 3.0 seconds indicate poor performance

FCP directly impacts the perceived speed of your website. A faster FCP indicates that users are seeing content sooner, leading to a better user experience and potentially lower bounce rates. It's one of the Core Web Vitals metrics that Google considers for assessing page experience and can affect your site's search engine rankings. Several factors influence FCP:

- **Server response time**: The time it takes for the server to respond to the initial request and start sending back resources

- **HTML structure**: Efficiently structured HTML can lead to quicker parsing and rendering by the browser

- **CSS and JavaScript**: Render-blocking CSS and JavaScript can delay FCP; optimizing and deferring non-critical CSS and JavaScript can improve FCP

- **Render path complexity**: The number of resources (images, scripts, and stylesheets) and their size impact how quickly the browser can render the first contentful paint

- **Network conditions**: The user's network speed affects how quickly resources can be fetched and rendered

Since the FCP metric is crucial for SEO, several techniques can be employed to improve it. Let's explore some of these methods in detail.

Techniques to improve FCP

Strategies to improve FCP often overlap with those used to enhance other Core Web Vitals such as **Largest Contentful Paint** (**LCP**), **Cumulative Layout Shift** (**CLS**), and **First Input Delay** (**FID**). However, each metric has specific factors and optimization techniques that target its unique aspects. Improving FCP typically involves optimizing the delivery and rendering of above-the-fold content. Here are some strategies:

- **Optimize critical rendering path**: Minimize critical resources by reducing the number and size of CSS and JavaScript files:

  - Prioritize *above-the-fold* content by inlining essential CSS for immediate rendering and deferring non-essential CSS to load asynchronously, enhancing page performance

  - Reduce server response times by using a **Content Delivery Network** (**CDN**) and optimize server configurations

- **Optimize HTML and CSS**: Remove unused CSS by eliminating any CSS not used for the initial view and **minify** CSS and HTML.

- **Optimize JavaScript**: You do this if you use minification and asynchronous loading:

  - Minimize the size of JS files and defer non-critical JS to load after the main content

  - Use asynchronous loading; load JavaScript files asynchronously to prevent them from blocking the rendering of the page

- **Reduce render-blocking resources**: Load critical resources first and ensure essential resources are loaded quickly using `defer` or `async` to non-critical CSS and JS, this ensures the initial content loads faster.

While improving FCP, LCP, CLS, and FID involves some overlapping strategies—such as optimizing JavaScript and CSS, reducing server response times, and prioritizing critical resources—each metric has specific optimizations that address its unique characteristics. By focusing on the specific needs of each metric, you can effectively enhance the overall user experience and performance of your website.

By implementing these techniques, you can significantly improve FCP, enhancing the perceived speed and usability of your website for users and potentially improving your site's search engine ranking.

LCP

LCP is a web performance metric that measures the time it takes for the largest content element on a webpage to become visible within the user's viewport.

It is a key component of Core Web Vitals and focuses on loading performance, aiming to reflect the user's perception of how quickly the main content of a page is rendered. The LCP metric considers elements such as the following:

- Images and image elements inside an `<img>` tag

- Background images loaded via CSS

- Block-level text elements such as `<p>` or `<h1>`

A good LCP score is 2.5 seconds or less, as illustrated in *Figure 10.3*. Scores exceeding 2.5 seconds indicate that improvements are needed.

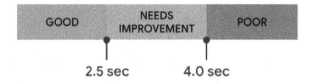

Figure 10.3 – LCP metrics

Improving LCP can involve optimizing server response times, leveraging caching, improving resource load times, and other techniques to ensure that the main content loads quickly for users.

Techniques to improve LCP

A fast LCP is crucial for providing a positive user experience, as it significantly impacts how quickly users perceive a page as fully loaded. To enhance LCP, various techniques can be employed to optimize load times and improve overall performance.

In this section, we'll explore some of these effective strategies to boost LCP, ensuring a smoother and more efficient user experience on your website. Let's examine each strategy in detail:

- **Optimize server response times**: Reduce server response times by using a fast web host, optimizing your database, and using a CDN. Implement server-side caching to deliver pages more quickly.

- **Use efficient coding practices**: Minify CSS, JavaScript, and HTML to reduce file sizes. Remove unnecessary CSS and JavaScript.

- **Optimize images and video**: Compress images without sacrificing quality using tools such as **ImageOptim** or **TinyPNG**. Serve images in next-gen formats such as **WebP**. Use responsive images to ensure that they are appropriately sized for different devices. **Lazy-load** offscreen images to defer loading until they are needed.

- **Improve client-side rendering**: Minimize JavaScript execution time by reducing the size of your JavaScript bundles. Defer non-critical JavaScript and use `async` or `defer` attributes to prevent it from blocking the rendering of the page.

- **Preload key resources**: Use the `<link rel="preload">` tag to preload important resources such as fonts, CSS, and images to speed up their loading.

- **Optimize fonts**: Use `font-display: swap` in your CSS to avoid invisible text during font loading. Limit the use of custom fonts and consider using system fonts.

- **Implement critical CSS**: Extract and inline critical CSS (the CSS required to render the above-the-fold content) directly into the HTML to speed up rendering.

- **Reduce third-party scripts**: Limit the use of third-party scripts as they can block the main thread and delay LCP. Use asynchronous loading for third-party scripts when possible.

- **Reduce blocking resources**: Identify and reduce render-blocking resources such as large CSS and JavaScript files. Use the `async` or `defer` attributes for JavaScript files that are not critical for initial rendering.

By applying these techniques, you can improve your website's LCP, resulting in a better user experience and potentially higher search engine rankings.

FID

FID is a web performance metric that measures the time it takes for a webpage to respond to the first user interaction. This interaction can be a click, tap, or key press. It does not measure the time taken to process the event itself, but rather the time the browser is busy doing other work and is unable to respond to the user. FID is part of the Core Web Vitals, emphasizing the importance of interactivity and user engagement.

A low FID means the page is quickly interactive, providing a smoother and more engaging user experience. A high FID indicates that the page feels sluggish, which can frustrate users and lead to higher bounce rates. Here's how we can gauge the FID scores:

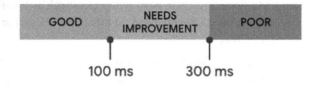

Figure 10.4 – FID metrics

In summary, FID measures the time it takes for a website to respond to the first user interaction, such as clicking a button or typing in a form field. A low FID is crucial for ensuring that users can interact with your website quickly and efficiently, contributing to a more responsive and engaging experience. Let's explore practical strategies to enhance FID, helping to create a more responsive and user-friendly web experience.

Techniques to improve FID

There are numerous techniques that can enhance FID, and it's important to highlight that many of these methods also contribute to improving other Core Web Vitals. Some of these strategies have already been discussed in relation to optimizing LCP:

- **Minimize main thread work**: Break up long tasks into smaller, asynchronous tasks to prevent the main thread from being blocked.

- **Optimize JavaScript execution**: Minify and compress JavaScript files to reduce their size. Remove unused JavaScript to avoid unnecessary code execution. Split large JavaScript bundles into smaller, more manageable chunks using code-splitting techniques.

- **Defer non-critical JavaScript**: Use the `async` or `defer` attributes on `<script>` tags to load JavaScript files without blocking the main thread. Defer the loading of non-essential scripts until after the page has been initially rendered.

- **Reduce the impact of third-party code**: Limit the use of third-party scripts, as they can introduce significant delays. Ensure that any third-party scripts are loaded asynchronously.

- **Optimize event listeners**: Avoid using passive event listeners where possible, as they can delay the browser's ability to respond to user interactions. Use event delegation to reduce the number of event listeners.

- **Use efficient coding practices**: Optimize CSS to ensure it doesn't block rendering. Inline critical CSS to speed up initial page rendering.

These techniques can enhance FID and contribute significantly to the overall SEO of the application. Next, we'll explore how to accurately measure FID.

CLS

CLS is a web performance metric that measures the visual stability of a webpage. Specifically, it quantifies how much visible content shifts in the viewport as the page loads and is interacted with. Visual stability is crucial for user experience. Unexpected shifts can lead to user frustration, accidental clicks, and a perception that the website is unreliable or poorly designed.

CLS is a part of the Core Web Vitals, emphasizing the importance of a visually stable user experience. A high CLS score indicates that users might be experiencing unexpected layout shifts, which can degrade their experience.

A layout shift occurs anytime a visible element changes its position from one frame to the next. The score is calculated based on the impact fraction and distance fraction of the shifting elements. Let's look at these fractions a little more here:

- **Impact fraction**: The percentage of the viewport that is affected by the shift

- **Distance fraction**: The distance that the elements have moved in the viewport

The CLS scores can be gauged as shown in the following figure:

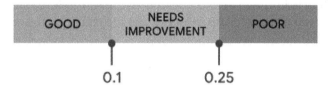

Figure 10.5 – The CLS metric

A score of less than 0.1 is considered good, indicating minimal unexpected layout shifts. Scores between 0.1 and 0.25 suggest room for improvement, where some elements may shift but the page is generally stable. A score exceeding 0.25 is deemed poor, indicating significant and disruptive layout shifts that can affect user experience negatively.

Achieving and maintaining a low CLS score is crucial for providing a seamless and user-friendly browsing experience.

Techniques to improve CLS

There are various techniques designed to improve CLS. It's noteworthy that many of these approaches also positively impact other Core Web Vitals:

- **Always include size attributes on images and videos**: Specify width and height attributes in the `<img>` and `<video>` tags to allocate space in the layout before the resources load.

- **Reserve space for ads and embeds**: Allocate fixed-size containers for advertisements and embedded content. Ensure that the allocated space is not smaller than the actual content.

- **Avoid inserting content above existing content**: Avoid inserting new content above existing content unless it is in response to a user action. For example, don't add new ads above the main content of the page after the page has loaded.

- **Use CSS to ensure stable layouts**: Use CSS to control animations and transitions. For example, prefer transform and opacity animations as they do not trigger layout changes. Ensure that any dynamic content changes do not push existing content out of place.

- **Preload important resources**: Use `<link rel="preload">` to preload critical resources, ensuring they are loaded quickly and reducing the likelihood of layout shifts.

- **Avoid layout jitter**: Ensure that interactive elements such as buttons and forms are rendered in their final state, minimizing changes after user interactions.

After implementing these techniques, it's crucial to accurately measure their impact on the CLS metric. In the next section, we will explore tools for measuring and analyzing CLS to assess and optimize its performance effectively.

How to measure SEO metrics

Understanding and optimizing key SEO metrics such as FCP, LCP, FID, and CLS is crucial for enhancing your website's performance and user experience. Various tools and resources can help you measure and analyze these essential metrics. Here are some of the most effective ones:

- **Lighthouse**: Lighthouse is integrated into Chrome DevTools and provides detailed insights into a range of performance metrics. It helps identify elements that slow down initial rendering, offering insights into FCP. It also provides comprehensive metrics to assess the time it takes for the LCP to become visible. Lighthouse measures FID, offering insights into how quickly a website responds to user interactions, such as clicks or keystrokes. For CLS, Lighthouse helps identify elements that cause unexpected layout shifts, ensuring better visual stability. This tool is a complete solution for performance analysis!

- **WebPageTest**: WebPageTest offers an in-depth analysis of loading performance, including detailed breakdowns and visualizations to help you understand loading sequences and bottlenecks, especially for FCP.

- **Web Vitals Extension**: The Web Vitals Extension is a Chrome extension that measures Core Web Vitals in real-time. It provides feedback on how quickly the largest visible content loads (LCP), insights into interaction responsiveness (FID), and real-time tracking of visual stability to help minimize layout shifts (CLS).

- **Google PageSpeed Insights**: Google PageSpeed Insights provides performance data and suggestions for improvement. It focuses on analyzing FCP and offers actionable insights to enhance initial load performance.

- **Chrome User Experience Report**: The Chrome User Experience Report aggregates real user data from Chrome users worldwide. It provides real-world data on how quickly large content elements become visible (LCP) and real user performance data to help you understand and optimize interaction latency (FID). For CLS, it offers insights into the visual stability of your site.

- **JavaScript APIs**: Using the `PerformanceObserver` interface, you can track FID programmatically, allowing for custom data collection and analysis. It also tracks CLS, enabling precise tracking of layout shifts.

By regularly monitoring and optimizing these Core Web Vitals using the tools and techniques mentioned, you can ensure your website delivers a smooth and responsive user experience. To see these measurements in action, let's analyze the SEO of a real application in the following exercise.

Exercise 10.2 – analyzing a real website with Lighthouse

Let's assess how the Packt Publishing website performs with Core Web Vitals using Google's Lighthouse:

1. First, for real-time analysis, visit `https://www.packtpub.com/en-br` using Google Chrome, open the Developer Tools (press *F12*), select **Lighthouse** from the top menu of the developer tools, and click on **Analyze Page Load.**

   Based on the Lighthouse analysis, Packt's website demonstrates strong performance in SEO metrics. *Figure 10.6* displays the first page of the Lighthouse report, summarizing the application's performance across various metrics.

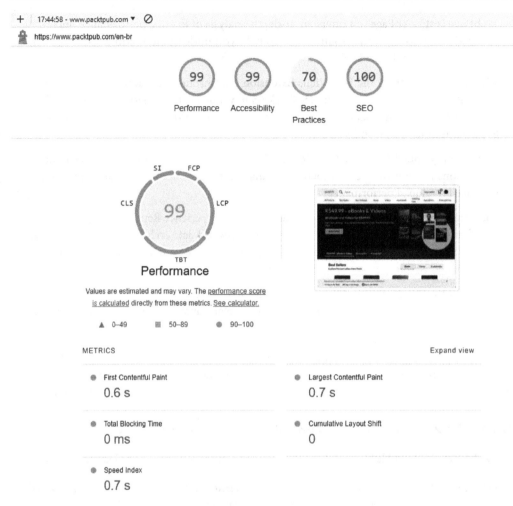

Figure 10.6 – Packt Publishing Lighthouse report on June 20, 2024

2. Now, let's go to the details. Lower on the same page is a section with more details. Here, you can delve into metrics and receive suggestions to enhance your website's performance. Click on FCP to see more about this metric, as shown in *Figure 10.7*:

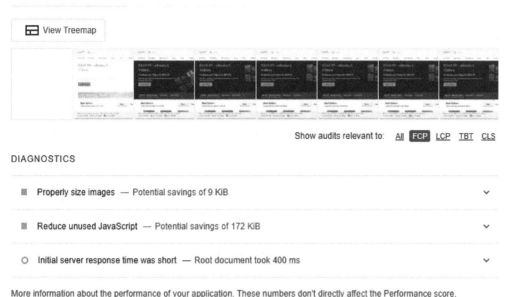

DIAGNOSTICS

▓ **Properly size images** — Potential savings of 9 KiB   ⌄

▓ **Reduce unused JavaScript** — Potential savings of 172 KiB   ⌄

○ **Initial server response time was short** — Root document took 400 ms ⌄

More information about the performance of your application. These numbers don't <u>directly affect</u> the Performance score.

Figure 10.7 – FCP diagnostics to improve performance generated by Google Lighthouse

3. Keep exploring Lighthouse! Use the filters to identify opportunities for improving your scores, and delve into the menus and resources the tool offers to gain deeper insights.

4. For further exploration, we recommend analyzing three additional websites of your choice using Lighthouse. Observe how their Core Web Vitals metrics compare and consider applying similar optimizations to achieve better results across different online platforms.

Utilizing Lighthouse offers crucial insights into optimizing website performance. By analyzing metrics and implementing recommended improvements, web developers can enhance user experience, improve site speed, and boost overall accessibility and SEO. Another effective SEO strategy is to minimize render-blocking resources, which we will explore in detail next.

Minimizing render-blocking resources

Render-blocking resources are CSS and JavaScript files that prevent the browser from rendering the page until they are fully loaded and executed. This delay is particularly noticeable for users on slower connections or less powerful devices.

Cumulative delays caused by multiple render-blocking resources can significantly extend the time it takes for users to interact with and perceive content on the page. This delay directly impacts SEO metrics such as FCP and can lead to a lower search engine ranking.

Understanding and minimizing render-blocking resources is crucial for optimizing web page performance, particularly in terms of improving loading times and user experience.

Before we go any further, let's look at the types of render-blocking resources:

- **CSS**: External stylesheets (linked via the `<link>` tag) and inline `<style>` tags can block rendering because the browser needs CSS to properly style the page

- **JavaScript**: Scripts (linked via the `<script>` tag) that are not marked as asynchronous (`async`) or deferred (`defer`) can block rendering because the browser must execute them before continuing to build the DOM

Since these resources can negatively impact SEO, developers use various techniques to minimize their effects and prevent or reduce render-blocking. Let's explore these methods in detail.

How to minimize render-blocking resources with CSS

To minimize render-blocking resources, one effective technique is to optimize CSS delivery. This involves two key strategies: identifying and inlining critical CSS and deferring non-critical CSS.

Critical CSS refers to the minimal CSS required to render the above-the-fold content, which is the portion of the webpage visible without scrolling. By inlining this essential CSS, you ensure that the necessary styling is applied quickly, enhancing the initial loading performance.

We can inline the minimal CSS required for the above-the-fold content directly into the HTML. This can be done within the `<head>` section of your HTML document:

```html
<!DOCTYPE html>
<html lang="en">
<head>
    <meta charset="UTF-8">
    <meta
        name="viewport"
        content="width=device-width,
                initial-scale=1.0"
    >
    <title>Optimized Page</title>
    <style>
        /* Critical CSS */
        body {
            font-family: Arial, sans-serif;
            background-color: #f4f4f4;
        }
        header {
            background-color: #333;
```

```
            color: #fff;
            padding: 1em;
            text-align: center;
        }
        /* Add more critical styles as needed */
    </style>
</head>
<body>
    <header>
        <h1>Welcome to My Website</h1>
    </header>
    <main>
        <!-- Main content -->
    </main>
</body>
</html>
```

For non-critical CSS, it is recommended to load these styles asynchronously after the initial page content has been rendered. This can be achieved by using JavaScript to dynamically load stylesheets or by employing the `<link rel="preload">` tag with the `as="style"` attribute.

These methods help ensure that the primary content loads faster, improving the overall user experience. Include the following in the `<head>` section of your HTML document to preload the non-critical CSS file and apply it once the page has loaded:

```
<!DOCTYPE html>
<html lang="en">
<head>
    <meta charset="UTF-8">
    <meta
        name="viewport"
        content="width=device-width,
                initial-scale=1.0"
    >
    <title>Optimized Page</title>
    <style>
        /* Critical CSS */
        body {
            font-family: Arial, sans-serif;
            background-color: #f4f4f4;
        }
        header {
            background-color: #333;
            color: #fff;
```

```
            padding: 1em;
            text-align: center;
        }
    </style>
    <link
        rel="preload"
        href="styles/non-critical.css"
        as="style"
        onload="this.rel='stylesheet'"
    >
    <noscript>
        <link
            rel="stylesheet"
            href="styles/non-critical.css"
        >
    </noscript>
</head>
<body>
    <header>
        <h1>Welcome to My Website</h1>
    </header>
    <main>
        <!-- Main content -->
    </main>
</body>
</html>
```

The `<noscript>` element is used to provide fallback content for users who have JavaScript disabled in their browsers. In the context of deferring non-critical CSS, the `<noscript>` block ensures that the non-critical CSS is still applied even if JavaScript is not available.

How to minimize render-blocking resources with JavaScript

To optimize JavaScript execution and improve webpage performance, several techniques can be employed:

- First, the `async` attribute should be used for external JavaScript files that are not essential for the initial rendering of the page. This allows the script to download asynchronously and execute as soon as it's ready, without blocking other resources. Use the `async` attribute for scripts that can run independently of other scripts and don't depend on the DOM being fully loaded:

```
<!DOCTYPE html>
<html lang="en">
<head>
    <meta charset="UTF-8">
```

```
    <meta
        name="viewport"
        content="width=device-width,
                initial-scale=1.0"
    >
    <title>Optimized Page</title>
</head>
<body>
    <!-- Page content -->

    <script
        src="path/to/non-essential-script.js"
        async
    ></script>
</body>
</html>
```

- Second, the `defer` attribute should be used for scripts that need to be executed in order but don't need to block the DOM construction; scripts with `defer` are executed after the HTML has been fully parsed. We can use the `defer` attribute for scripts that need to be executed in order and after the HTML has been parsed:

```
<!DOCTYPE html>
<html lang="en">
<head>
    <meta charset="UTF-8">
    <meta
        name="viewport"
        content="width=device-width,
                initial-scale=1.0"
    >
    <title>Optimized Page</title>
    <script src="path/to/script1.js" defer></script>
    <script src="path/to/script2.js" defer></script>
</head>
<body>
    <!-- Page content -->
</body>
</html>
```

- Additionally, loading JavaScript conditionally can further enhance performance. This includes lazy loading, which delays the loading of non-critical JavaScript until it is needed, such as after a user action or when it becomes visible. We can lazy load non-critical JavaScript to improve initial load performance:

```html
<!DOCTYPE html>
<html lang="en">
<head>
    <meta charset="UTF-8">
    <meta
        name="viewport"
        content="width=device-width,
                initial-scale=1.0"
    >
    <title>Optimized Page</title>
</head>
<body>
    <!-- Page content -->

    <script>
        document.addEventListener(
            'DOMContentLoaded',
            function() {
                var lazyScript =
                    document.createElement('script');
                lazyScript.src =
                    'path/to/non-critical-script.js';
                document.body.appendChild(lazyScript);
            }
        );
    </script>
</body>
</html>
```

- On-demand loading involves using dynamic script loading techniques, such as `document.createElement('script')`, to load scripts only when they are necessary. We should load scripts only when they are necessary using dynamic script creation:

```html
<!DOCTYPE html>
<html lang="en">
<head>
    <meta charset="UTF-8">
    <meta
        name="viewport"
```

```
            content="width=device-width,
                    initial-scale=1.0"
    >
    <title>Optimized Page</title>
</head>
<body>
    <!-- Page content -->
    <button id="loadScriptButton">Load Script</button>

    <script>
        document.getElementById(
            'loadScriptButton'
        ).addEventListener(
            'click',
            function() {
                var script =
                    document.createElement('script');
                script.src =
                    'path/to/conditional-script.js';
                document.body.appendChild(script);
            }
        );
    </script>
</body>
</html>
```

By using these techniques, you can significantly minimize render-blocking resources of your webpage by ensuring that JavaScript is loaded and executed in the most efficient manner.

By understanding how render-blocking resources affect page load times and employing techniques to minimize their impact, you can significantly improve the perceived speed and user experience of your website.

Prioritize loading critical resources first, defer or asynchronously load non-critical resources, and regularly monitor and optimize your site's performance to ensure fast and efficient rendering. These practices not only enhance user satisfaction but also contribute positively to SEO and conversion rates.

Optimizing images for SEO

By optimizing images for SEO, you not only enhance the aesthetic appeal and accessibility of your website but also contribute to its overall performance, visibility, and user satisfaction.

Optimizing images for SEO involves several practices that help ensure images contribute positively to a website's search engine ranking and overall user experience. Here are some key strategies:

- **Choose the right file format**:

  - **JPEG**: Best for photographs and images with many colors, JPEG offers a good balance between quality and file size

  - **PNG**: Best for images that require transparency and for simpler images with fewer colors

  - **WebP**: A modern format that provides superior compression for both lossless and lossy images, supported by most browsers

  - **SVG**: Best for logos, icons, and simple graphics; it's a vector format, so it's scalable without loss of quality

- **Compress images**: Use tools such as *TinyPNG*, *ImageOptim*, or *ShortPixel* to reduce image file sizes without compromising quality. Ensure that images are appropriately compressed to decrease load times, improving page speed, which is a ranking factor.

- **Use descriptive file names**: Rename image files to include relevant keywords before uploading them. Instead of `IMG_1234.jpg`, use `blue-widget-for-sale.jpg`. Use hyphens to separate words (not underscores).

- **Implement responsive images**: Use the `srcset` attribute to serve different image sizes based on the user's device, ensuring images are not larger than necessary. Implement the `<picture>` element to offer different image formats.

- **Use lazy loading**: Implement lazy loading to defer the loading of offscreen images until the user scrolls near them. This improves initial page load time. Use the `loading="lazy"` attribute on `<img>` tags or JavaScript libraries for lazy loading.

- **Ensure images are crawlable**: Make sure images are not blocked by `robots.txt`. Use image sitemaps to help search engines discover and index images on your site.

- **Add structured data**: Use structured data (*schema.org*) to mark up images with additional information such as product details, recipes, and more. This can enhance search results with rich snippets and improve visibility.

- **Optimize image dimensions**: Serve images at the correct dimensions to avoid browser scaling. This ensures that images look sharp and load faster. Properly sizing images for the web involves optimizing images so they are displayed at the correct dimensions and file size, which helps improve the overall user experience.

  Use CSS to control image display and avoid inline styles that can override responsive behavior:

  - **HTML/CSS layout**: Determine the maximum display size of the image on your webpage using HTML/CSS

- **Responsive design**: Consider responsive design principles to ensure images scale appropriately on different devices and screen sizes

Here's an example of a properly sized image with responsive design:

```
<img
    src="image.jpg"
    alt="Description of the image"
    width="800"
    height="600"
>
<!--Example of using srcset for responsive images -->
<img src="image.jpg"
    srcset="image-400.jpg 400w,
            image-800.jpg 800w,
            image-1200.jpg 1200w"
    sizes="(max-width: 600px) 400px,
           (max-width: 1000px) 800px,
           1200px"
    alt="Description of the image">
```

- **Leverage CDNs**: Use CDNs to distribute image content globally, reducing latency and improving load times for users around the world.

- **Use caption and context**: Provide captions and context for images. Search engines use surrounding text to understand the image content. Well-described images in context can rank better in image search.

- **Regularly audit and update**: Regularly audit your site's images for performance and SEO improvements. Update alt text, compress images, and check for broken links or outdated content.

Properly optimizing images in HTML is essential for enhancing both accessibility and SEO. Here is an example of HTML code demonstrating how to effectively use an optimized image:

```
<picture>
  <source srcset="image.webp" type="image/webp">
  <source srcset="image.jpg" type="image/jpeg">
  <img
    src="image.jpg"
    alt="Description of the image"
    loading="lazy" width="600"
    height="400"
  >
</picture>
```

We have established that optimizing images is crucial for improving website performance, user experience, and SEO. Here are some tools and resources that can help streamline the image optimization process:

- **Image compression**: TinyPNG, JPEG-Optimizer, and ImageOptim

- **Responsive images**: Cloudinary and Imgix

- **Lazy loading**: LazySizes and the native HTML `loading="lazy"` attribute

These tools and resources cater to different aspects of image optimization, from compression and resizing to accessibility and SEO enhancement. By leveraging these tools effectively, web developers and content creators can ensure that their images contribute positively to overall website performance and user satisfaction.

By following these practices, you can ensure that your images contribute to a better user experience and improved SEO performance, helping your website rank higher in search results and perform better overall. Next, let's examine the best practices for using the alt tag for images in accordance with SEO techniques.

Using alt tags for images

We need to write descriptive alt text that accurately describes the image content and includes relevant keywords. Alt text helps search engines understand the context of an image and improves accessibility for screen readers. Next, we'll examine both effective and ineffective examples of alt tags.

Examples of good and bad alt tags

Good alt tags are descriptive and concise. They provide context that is relevant to the image and the content surrounding it. They also include keywords where appropriate but avoid keyword stuffing.

Bad alt tags are either too vague, too detailed, irrelevant, or stuffed with keywords. They fail to provide meaningful context or accessibility.

Let's take a look at the following figure:

Figure 10.8 – Slice of cheesecake covered with cherry jelly.

Next let's look at two examples of alt tags here:

- **Good alt tag**:

```
<img
  src="cheesecake-cherry-jelly.jpg"
  alt="Slice of cheesecake covered with cherry jelly
      on a white plate with a fork on a table with a
      white towel"
>
```

This description is specific and detailed, including relevant keywords such as *"cheesecake,"* *"cherry jelly,"* *"white plate,"* *"fork,"* and *"table with a white towel."* This helps search engines understand the content of the image, improving SEO, and also provides a clear description for users who rely on screen readers.

- **Bad alt tag**:

```
<img src="cheesecake-cherry-jelly.jpg" alt="Dessert">
```

This description is vague and generic. It lacks specific details and relevant keywords, making it less useful for SEO and less informative for users with visual impairments.

Let's take another example within this figure:

Figure 10.9 – Jack Russell Terrier puppy looking at the camera.

Once again, let's look at two examples of alt tags here:

- **Good alt tag**:

```
<img
  src="jack-russell-puppy.jpg"
  alt="Curious Jack Russell Terrier puppy looking at
      the camera in a green field"
>
```

This description is specific and detailed, including relevant keywords such as "*Jack Russell Terrier*," "*puppy*," "*looking at the camera*," and "*green field*." This helps search engines understand the content of the image, improving SEO, and provides a clear and concise description for users relying on screen readers.

- **Bad alt tag**:

```
<img
  src="jack-russell-puppy.jpg"
  alt="Puppy dog Jack Russell Terrier cute puppy green
      field curious puppy looking camera puppy dog
      terrier field green"
>
```

This description is overly stuffed with keywords, making it look spammy and reducing its effectiveness for SEO. It lacks clarity and readability, which can be detrimental to both search engines and users who rely on screen readers.

In summary, writing effective alt tags is crucial for image search optimization. Crafting good alt tags requires practice and an understanding of how search engines handle images. Let's delve into this topic further.

Best practices for writing effective alt tags

Effective alt tags are crucial for accessibility and SEO. Here are some guidelines for writing better alt tags:

- They should be **descriptive** and **specific**, accurately describing the image content and function without unnecessary detail.

- Use **relevant keywords naturally**, ensuring the alt text is **concise** and **contextually relevant** to the surrounding content

- Avoid **redundancy** and phrases such as *image of* or *picture of*

- For functional images, describe the function they serve, while decorative images should have an empty (`alt=""`) alt attribute

- Regularly **review and update alt text** to maintain accuracy and usefulness.

Following these practices ensures alt tags contribute positively to both accessibility for users with disabilities and SEO efforts on your website.

Another important element impacting SEO is the DOM. The DOM acts as a map or tree that represents the structure of a web page, including all nodes and their relationships. Keeping the DOM size as small as possible is crucial for optimizing performance, and we'll explore how to achieve this in the following section.

Avoiding excessive DOM size

The DOM represents the structure of HTML elements on a webpage and directly impacts how quickly the browser can render and interact with the page. DOM size refers to the total number of elements (HTML tags), including nodes and their relationships (parent-child or sibling), styles, and attributes, that make up the structure of a webpage.

Each element in the DOM requires memory and processing time for rendering and scripting operations. A large DOM size can lead to slower performance, increased memory usage, and potential rendering issues. This is why avoiding excessive DOM size is crucial for optimizing web page performance and ensuring a smooth user experience.

Techniques to Avoid Excessive DOM Size

To avoid excessive DOM size, several techniques can be employed:

- **Optimize your HTML structure by simplifying markup**. Use semantic HTML and avoid unnecessary nesting of elements, keeping the structure as flat as possible. Remove redundant elements, attributes, and inline styles.

- **Limit dynamic content by implementing lazy loading**, which loads content dynamically only when needed. Also, use techniques such as pagination or infinite scroll to progressively load content as the user scrolls, rather than all at once.

- **Efficient use of CSS and JavaScript is also crucial**. Minimize CSS complexity by reducing the number of CSS rules and selectors, and avoid excessive styles that apply to many elements. Optimize JavaScript by refactoring code to minimize DOM manipulation operations and improve performance.

- Additionally, optimize rendering performance by **prioritizing the loading and rendering of critical resources** (CSS and JavaScript) necessary for above-the-fold content. Use asynchronous loading (the `async` attribute) or defer execution (the `defer` attribute) for non-critical scripts to improve initial rendering speed.

- Regularly **monitor and audit DOM size** using browser developer tools such as *Chrome DevTools* to inspect and analyze DOM size and structure, identifying and addressing elements or scripts that contribute excessively to DOM size and performance bottlenecks.

- Finally, consider the **impact of JavaScript frameworks and libraries** (such as *React*, *Angular*, or *Vue.js*) on DOM size and performance, optimizing their usage and evaluating alternatives if necessary.

Here's an example of a complex DOM structure:

```
<div class="container">
    <div class="header">
        <h1>Welcome to My Website</h1>
        <p class="intro">
            This is the best place to find great content.
        </p>
    </div>
    <div class="main-content">
        <div class="section">
            <h2>Section 1</h2>
            <div class="section-content">
                <p>This is some text for section 1.</p>
                <div class="extra-info">
                    <p>Additional info 1</p>
```

```
                    </div>
                </div>
            </div>
            <div class="section">
                <h2>Section 2</h2>
                <div class="section-content">
                    <p>This is some text for section 2.</p>
                    <div class="extra-info">
                        <p>Additional info 2</p>
                    </div>
                </div>
            </div>
        </div>
        <div class="footer">
            <p>Footer content here.</p>
        </div>
    </div>
```

The following is the same DOM in a simplified version:

```
<div class="container">
    <header>
        <h1>Welcome to My Website</h1>
        <p>
            This is the best place to find great content.
        </p>
    </header>
    <main>
        <section>
            <h2>Section 1</h2>
            <p>This is some text for section 1.</p>
            <p>Additional info 1</p>
        </section>
        <section>
            <h2>Section 2</h2>
            <p>This is some text for section 2.</p>
            <p>Additional info 2</p>
        </section>
    </main>
    <footer>
        <p>Footer content here.</p>
    </footer>
</div>
```

In this code, we replaced the generic `<div>` elements with semantic elements. Also, we removed unnecessary `<div>` wrappers around text and sections. By simplifying the DOM structure in this way, the HTML becomes more readable, maintainable, and efficient.

For more insights into how DOM size affects user experience and strategies to manage it, check out Jeremy Wagner's article on web.dev (`https://web.dev/articles/dom-size-and-interactivity`). Additionally, the **Lighthouse documentation** offers a comprehensive guide on reducing DOM size:

`https://developer.chrome.com/docs/lighthouse/performance/dom-size`

By prioritizing simplicity, efficient resource loading, and responsiveness, you can ensure a faster and smoother user experience across various devices and network conditions.

Summary

In this chapter, we delved into several critical aspects that shape the performance and visibility of websites. We explored the importance of Core Web Vitals such as LCP, FID, and CLS, emphasizing their impact on user experience and search engine rankings. Additionally, we discussed strategies for optimizing images, improving FCP, and minimizing render-blocking resources and excessive DOM size. By focusing on these fundamental elements, webmasters and developers can enhance website speed, accessibility, and overall SEO performance, ultimately delivering a superior user experience and achieving higher search engine rankings.

In the next chapter, we will explore preprocessors and tooling that play a crucial role in efficient web development practices. Preprocessors such as **Sass** and **Less** enable developers to write more maintainable and organized CSS, utilizing features such as variables, mixins, and nesting. By leveraging these preprocessors and tools effectively, developers can boost productivity, code quality, and ultimately, the performance and SEO of their websites.

11

Preprocessors and Tooling for Efficient Development

CSS **preprocessors** such as **Sass** and **Less** have evolved significantly since their inception, transforming how developers approach styling on the web.

Sass, one of the earliest preprocessors, was created in 2006 by Hampton Catlin and developed further by Natalie Weizenbaum. It introduced a new way of writing CSS with features such as variables and mixins, which were revolutionary at the time. Less followed soon after, offering similar capabilities with a syntax closer to traditional CSS.

By enabling the use of programming concepts in CSS, preprocessors have made it easier to maintain large code bases, adapt to design changes, and promote code reuse. As a result, they've become essential tools for modern web development, helping teams deliver high-quality projects more efficiently.

Speaking of efficiency, **Gulp** and **Webpack** are essential tools in modern web development that streamline project workflows and optimize builds. Gulp, a task runner introduced in 2013, automates repetitive tasks such as minification and live reloading, making development more efficient. Webpack, a module bundler released in 2012, allows developers to bundle JavaScript modules and other assets, optimizing them for production.

This chapter explores these tools, equipping you with the skills needed to streamline your workflow and create more efficient and maintainable web applications.

In this chapter, we will cover the following topics:

- Understanding CSS preprocessors – Sass and Less
- Enhancing CSS with preprocessors
- Mastering build tools such as Gulp and Webpack to automate tasks such as compilation and optimization
- Optimizing web performance to ensure faster loading times and better user experience

This chapter aims to provide a practical foundation for using preprocessors and tooling effectively, empowering you to elevate your web development skills and deliver exceptional results in your projects.

Technical requirements

Before diving into the world of preprocessors and tooling, you must set up your development environment with the necessary technical requirements.

First, you should install **Visual Studio Code** (**VS Code**) or any other **integrated development environment** (**IDE**) of your choice. Additionally, ensure you have Node.js installed on your machine as it serves as the backbone for running various tools and packages you'll encounter.

Node.js is crucial for managing dependencies and executing JavaScript-based build tools. Other software, such as specific packages and tools such as Less and Gulp, will be installed during the practice sessions outlined in this chapter.

This setup will prepare you to explore and utilize the powerful capabilities of preprocessors and modern web development tooling.

The code files for this chapter can be found at `https://packt.link/hrpje`.

Understanding CSS preprocessors – Sass and Less

Imagine you're starting a web project, and you want to use CSS to style your website. Typically, CSS is great, but it can become repetitive and hard to manage as your project becomes larger. This is where CSS preprocessors such as Sass or Less come into play.

CSS preprocessors are tools that extend the capabilities of standard CSS, allowing you to write code in a more efficient, maintainable, and scalable way. The most commonly used CSS preprocessors are Sass and Less.

Using CSS preprocessors can significantly improve your workflow by making your code more concise, easier to read, and simpler to maintain. They add a layer of abstraction over the standard CSS, which can lead to more powerful and flexible stylesheets.

Let's take a look at this example:

```
/* styles.css */

body {
    font-family: Arial, sans-serif;
    color: #333;
}
```

```
h1 {
    font-size: 24px;
    color: #FF5733;
}

.container {
    width: 100%;
    margin: 0 auto;
}
```

The preceding snippet shows a common CSS file structure. Now, with Sass, you can make this more efficient and maintainable:

```
/* styles.scss */
$primary-color: #333;
$accent-color: #FF5733;

body {
    font-family: Arial, sans-serif;
    color: $primary-color;
}

h1 {
    font-size: 24px;
    color: $accent-color;
}

.container {
    width: 100%;
    margin: 0 auto;
}
```

Notice the use of variables ($primary-color, $accent-color) and nesting (body, h1, .container). Preprocessors such as Sass allow you to write cleaner code with reusable components and then compile it into regular CSS that browsers understand.

While Sass and Less both offer these advantages, they use slightly different syntaxes. Here's how the same code looks when written in Less:

```
/* styles.less */

@font-family: Arial, sans-serif;
@text-color: #333;
@header-color: #FF5733;
@container-width: 100%;
```

```
body {
    font-family: @font-family;
    color: @text-color;
}

h1 {
    font-size: 24px;
    color: @header-color;
}

.container {
    width: @container-width;
    margin: 0 auto;
}
```

Now that you've got an idea of how these preprocessors work, we can dive into their features, such as variables, mixins, and nesting, and explore each in detail as we move forward.

The benefits of features such as variables, mixins, and nesting

By introducing variables, nesting, mixins, and functions, preprocessors streamline the CSS workflow, making it more efficient and maintainable. Here's a breakdown of these features:

- **Variables**: With preprocessors, you can define variables to store values such as colors, fonts, or any CSS value that you need to reuse throughout your stylesheet. In this way, you can easily maintain consistency and quickly update your styles by changing the variable value.

- **Nesting**: Preprocessors allow you to nest your CSS selectors so that they follow the same visual hierarchy of your HTML. This makes your CSS more readable and organized.

- **Mixins**: These are reusable chunks of code that you can define once and include in multiple places. They are particularly useful for applying sets of styles that are used repeatedly.

- **Partials and imports**: Preprocessors let you split your CSS into smaller, manageable files and then combine them into a single CSS file. This modular approach helps in organizing your code better and makes it easier to maintain.

- **Functions and operations**: Preprocessors offer built-in functions and allow you to perform operations, such as darkening a color or performing calculations, directly in your CSS.

- **Extends**: These let you share a set of CSS properties from one selector to another, reducing redundancy and ensuring consistency.

CSS preprocessors work by taking the enhanced syntax and features you write in the preprocessor's language and compiling it into standard CSS that browsers can understand. Here's a step-by-step explanation of how they work:

1. **Write preprocessor code**: You start by writing your styles using the syntax and features of a CSS preprocessor such as Sass or Less. This might include variables, nested rules, mixins, functions, and other advanced features.

2. **Compile to CSS**: The preprocessor tool takes this code and compiles it into standard CSS. This compilation process translates the preprocessor-specific syntax into regular CSS rules. This process can be done using an IDE, a command-line tool, or directly in the browser, depending on the preprocessor you're using.

3. **Use the compiled CSS**: The resulting CSS file is what you include in your HTML files. Browsers can only interpret standard CSS, so they don't directly interact with the preprocessor code.

Each preprocessor has its particularities and differences in usage. Let's delve deeper into these differences to understand which one is better suited for our project.

The differences between Sass and Less

Choosing between Sass and Less often comes down to the specific needs of your project and your personal or team's preference. Both preprocessors significantly improve the CSS development process by adding useful features and improving maintainability.

While Sass is more feature-rich and powerful, making it suitable for larger projects and teams that need advanced functionality, Less is simpler and easier to get started with, making it a good choice for smaller projects or developers who prefer minimal setup:

- **Less** stands for *Leaner Style Sheets* and provides a dynamic syntax for coding CSS with logical patterns. It enhances cross-browser compatibility by optimizing the generated CSS. With Less, users can create reusable properties, and it is based on JavaScript.

- **SASS** stands for *Syntactically Awesome Style Sheets*. This preprocessor, implemented in Ruby, actively reports syntax errors, making it easier to catch mistakes. Like Less, Sass supports user customization, including variables, nesting, and mixins.

Here's a table highlighting the differences between Less and Sass in terms of variables, nesting, particularities, and best uses:

Feature	LESS	Sass
Variables	Variables are defined using the @ symbol (for example, `@color: #4D926F;`).	Variables are defined using the $ symbol (for example, `$color: #4D926F;`).

Feature	LESS	Sass
Nesting	Supports nesting similar to Sass but uses slightly different syntax.	Supports nesting with clean and intuitive syntax, allowing deep nesting.
Particularities	Less uses JavaScript for compilation, allowing it to be run in the browser.	Sass has two syntaxes: SCSS (more CSS-like) and Sass (indentation-based). It requires Ruby or other implementations for compilation.
Best used for	Projects where simplicity and ease of integration with JavaScript are important.	Projects needing advanced features, larger teams, or where a Ruby-based workflow is preferred.

Table 11.1 – Differences between Less and Sass

Now, let's consider the key differences between Sass and Less.

Syntax

Sass offers two syntaxes: the older indented syntax (Sass) and the newer SCSS syntax. The SCSS syntax is more similar to standard CSS, making it easier for those who are already familiar with CSS. Here's an example of Sass code:

```
$primary-color: #333;
.navbar {
  color: $primary-color;
  .nav-item {
    padding: 10px;
  }
}
```

The Less syntax is very similar to CSS, with a few additions for variables and nested rules. Here's an example:

```
@primary-color: #333;
.navbar {
  color: @primary-color;
  .nav-item {
    padding: 10px;
  }
}
```

Variables

Sass variables are defined using the $ symbol:

```
$font-stack: Helvetica, sans-serif;
$primary-color: #333;
```

Less variables are defined using the @ symbol:

```
@font-stack: Helvetica, sans-serif;
@primary-color: #333;
```

Functions and mixins

Sass provides powerful built-in functions and mixins. You can create complex functions and control structures such as loops and conditionals, as shown here:

```
@mixin border-radius($radius) {
  -webkit-border-radius: $radius;
    -moz-border-radius: $radius;
     -ms-border-radius: $radius;
        border-radius: $radius;
}
.box { @include border-radius(10px); }
```

Less also supports mixins and functions, but the syntax is a bit different:

```
.border-radius(@radius) {
  -webkit-border-radius: @radius;
    -moz-border-radius: @radius;
     -ms-border-radius: @radius;
        border-radius: @radius;
}
.box { .border-radius(10px); }
```

Nesting

Both Sass and Less allow you to nest your CSS selectors in a way that mirrors the HTML structure, making your stylesheets more readable and organized, with few syntax differences. Here's an example of nesting in Sass:

```
.navbar {
  .nav-item {
    color: $primary-color;
  }
}
```

Here's the same code in Less:

```less
.navbar {
  .nav-item {
    color: @primary-color;
  }
}
```

Partials and imports

Sass uses @import to include partial files. These partials are usually named with a leading underscore:

```scss
@import 'variables';
@import 'mixins';
```

Less also uses @import, but it doesn't have a naming convention for partials:

```less
@import 'variables.less';
@import 'mixins.less';
```

Compiling

Sass is typically compiled using command-line tools, build systems such as Gulp or Webpack, or IDE plugins.

Less can be compiled in the browser using JavaScript, as well as through command-line tools and build systems.

Community and ecosystem

Sass has a larger community and a more extensive ecosystem with many frameworks and libraries built around it (for example, Compass, Bourbon, and others).

Less also has a robust community, but it's somewhat less extensive compared to Sass.

Both preprocessors are valuable tools, and learning either will improve your CSS workflow. When it comes to deciding which CSS preprocessor is easier to learn, Less tends to be more beginner-friendly. Once you're completely comfortable with the basic concepts of preprocessors, you might consider exploring Sass to take advantage of its more advanced features and capabilities.

Less syntax is more straightforward and closely resembles standard CSS, making it easier for beginners to pick up. Also, Less can be compiled in the browser, which simplifies the setup process and allows beginners to see changes immediately without needing to configure additional tools. In the next section, we'll learn how to set up and do some exercises with Less.

Setting up and using Less in a project

The following section will take you through a step-by-step guide to get you started with Less. Once we've installed Less and its associated packages, we'll create and compile our first Less project.

Installing Node.js and LESS

To compile Less files into CSS, we need to install **Node.js**, **Node Package Manager** (**npm**), and the **npm LESS package** to run the compilation script.

First, we'll install Node.js. By navigating to `https://nodejs.org/en` in your web browser, you can find the download links for the Node.js installation package for your operating system, as illustrated in the following screenshot:

Figure 11.1 – NodeJS home page with a download link

After installing Node.js and npm, open your Terminal window (or Command Prompt on Windows). At the prompt, run the `node -v` and `npm -v` command. Both commands should return a version number, as illustrated in the following figure:

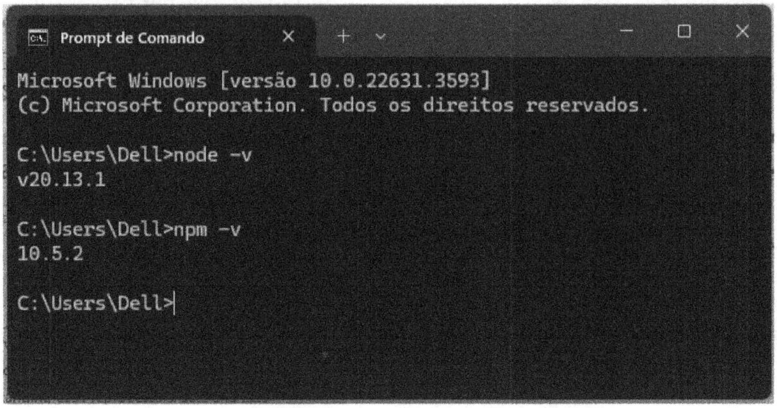

Figure 11.2 – Node and npm versions shown in the Terminal

This means that Node.js and npm are installed on your machine and ready to use. Now, let's install Less. Once again, open your Terminal and type the following:

```
npm install -g less
```

This command will install Less *globally* on your computer, allowing you to use it across all your projects without needing to install it each time. Once installed globally, you can use the `lessc` command in any project to compile Less files to CSS:

```
lessc styles.less styles.css
```

If installed correctly, your Terminal should look as follows:

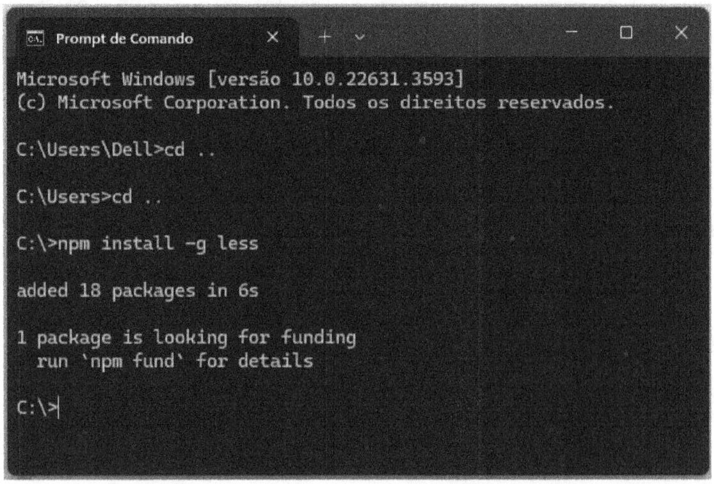

Figure 11.3 – Successful installation of Less globally

Global installation is convenient for quickly using Less across multiple projects without repeated installations.

Exercise 11.01 – Creating and compiling our first Less project

Now that we've installed Node.js, npm, and Less, let's create a new project using Less and HTML. Follow these steps:

1. Create a folder on your computer and name it `less-project`.

2. Open this folder in VS Code or your preferred IDE.

3. Inside the folder, create two files: `index.html` and `styles.less`.

4. In `index.html`, add the following code:

```
<!DOCTYPE html>
<html lang="en">
<head>
    <meta charset="UTF-8">
    <meta
        name="viewport"
        content="width=device-width,
        initial-scale=1.0">
    <title>LESS Project</title>
    <link rel="stylesheet" href="./styles.less">
</head>
<body>
    <h1>Welcome to LESS</h1>
    <button class="btn-primary">Click Me</button>
</body>
</html>
```

5. In `styles.less`, add the following code:

```
@primary-color: #3498db;
@font-size: 16px;

body {
    font-family: Arial, sans-serif;
    font-size: @font-size;
    color: @primary-color;
}

h1 {
    color: lighten(@primary-color, 20%);
```

```
    }

    .btn-primary {
        background-color: @primary-color;
        border: none;
        padding: 10px;
        color: white;
        &:hover {
            background-color: darken(@primary-color, 10%);
        }
    }
}
```

6. Now, let's convert Less into CSS. Ensure you're in the project's folder, then execute the following command in your Terminal to compile `styles.less` to `styles.css`:

```
lessc styles.less styles/styles.css
```

Note that it has created a folder named `'styles'` containing a `styles.css` file. If you navigate to this folder, you'll find the compiled CSS generated by our Less code, as shown here:

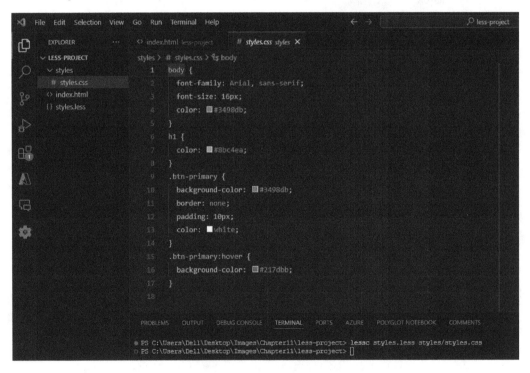

Figure 11.4 – Compiled CSS file generated with Less

7. After compiling, it's important to replace the CSS `<link>` tag so that it points to the new CSS-generated file. Go to the HTML file and replace `<link rel="stylesheet" href="./styles.less">` with `<link rel="stylesheet" href="./styles/styles.css">`:

```
<!DOCTYPE html>
<html lang="en">
<head>
    <meta charset="UTF-8">
    <meta
        name="viewport"
        content="width=device-width,
        initial-scale=1.0">
    <title>LESS Project</title>
    <link rel="stylesheet" href="./styles/styles.css">
</head>
<body>
    <h1>Welcome to LESS</h1>
    <button class="btn-primary">Click Me</button>
</body>
</html>
```

Congratulations! You've completed your first Less project. While it may seem like overkill for a simple project, Less offers powerful features that save time and ensure consistency and performance in your projects. If everything went smoothly, your output should look as follows:

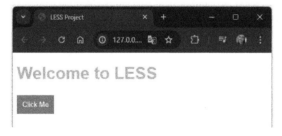

Figure 11.5 – Styled output made with Less

In this exercise, we created some Less code that necessitates a better understanding of its syntax. Let's delve into what each part of the code accomplishes.

Variables

Less allows you to define variables (`@primary-color`, `@font-size`) to store reusable values, making it easier to maintain and update styles across your project:

```
@primary-color: #3498db;
@font-size: 16px;
```

Let's take a closer look:

- @primary-color: Defines a variable for the primary color to be used throughout the stylesheet, set to the hex color #3498db

- @font-size: Defines a variable for the base font size, set to 16px

Nested selectors and functions

Less supports nesting of selectors (& refers to the parent selector), which helps in organizing CSS rules and improves readability:

```
body {
    font-family: Arial, sans-serif;
    font-size: @font-size;
    color: @primary-color;
}
```

Let's take a closer look:

- body: Styles the body element

- font-family: Sets the font family to Arial or sans-serif as a fallback

- font-size: Uses the value of the @font-size variable

- color: Sets the text color to the @primary-color variable

Less also provides built-in functions such as lighten() and darken() to manipulate colors dynamically, enhancing flexibility in styling:

```
h1 {
    color: lighten(@primary-color, 20%);
}
```

Let's take a closer look:

- h1: Styles all h1 headings

- color: lighten(): This is a Less function that lightens @primary-color by 20%, producing a lighter shade of the primary color when it's applied to h1 elements

In the following example, we use the nested &:hover selector to define the styles that will be applied to the .btn-primary class when the element is in a hover state:

```
.btn-primary {
    background-color: @primary-color;
    border: none;
    padding: 10px;
```

```
    color: white;
    &:hover {
        background-color: darken(@primary-color, 10%);
    }
}
```

Let's take a closer look at this code:

- `.btn-primary`: Styles all elements with the `.btn-primary` class. Typically, buttons are styled as primary action buttons.

- `background-color`: Sets the background color to `@primary-color`.

- `border`: Removes the border.

- `padding`: Adds 10 pixels of padding.

- `color`: Sets the text color to white.

- `&:hover`: Uses a nested selector to style the button when hovered over.

- `background-color`: The `darken()` function darkens `@primary-color` by 10% when the button is hovered over.

By utilizing variables, nested selectors, and functions effectively, this Less code demonstrates how to create styles that are modular and maintainable. This approach provides greater control over styling elements and enhances efficiency in CSS development. It enables developers to write cleaner, more organized code that is easier to maintain and update as project requirements evolve.

Five ways to learn to code with Less

Understand basic CSS: Ensure you have a solid understanding of CSS fundamentals – Less builds upon CSS concepts.

Explore the Less documentation: Visit the official website (`www.lesscss.org`) and explore the documentation, which includes comprehensive guides, examples, and references.

Practice: Start by converting existing CSS files into Less and experiment with variables, nesting, and functions. Practice using Less in small projects to familiarize yourself with its features.

Tool integration: Integrate Less into your development workflow by using task runners such as Gulp or Webpack, which can automate the compilation process and enhance productivity.

Stay updated: Keep up with new features and updates in Less to leverage its full potential and stay current with best practices in CSS preprocessing.

By following these steps and continuously practicing and experimenting with Less, you'll gradually become proficient in using it to streamline your CSS development process and create more efficient and maintainable stylesheets.

Now that we've set up our first Less project, we can delve deeper into the world of preprocessors by exploring a powerful feature: automation. In the next section, we'll learn how to automate the compilation process so that we can streamline our workflow.

Exercise 11.02 – automating compilation

When working with Less or any other preprocessor, you'll soon find yourself tired of manually recompiling your code every time you make even a minor change to the Less file just to see the updated result in your browser.

Imagine if there were a way to automate this process, allowing you to focus your energy on creating amazing user experiences rather than worrying about Less compilation.

To achieve this automation, in this exercise, we'll integrate less-watch-compiler into our project to ensure that our stylesheets are automatically compiled whenever changes are made:

1. **Install less-watch-compiler.**

    First, install less-watch-compiler globally on your machine. Open your Terminal or Command Prompt, navigate to the root folder (for example, C: for Windows users), and type the following command:

    ```
    npm install -g less-watch-compiler
    ```

    Your prompt should look like this:

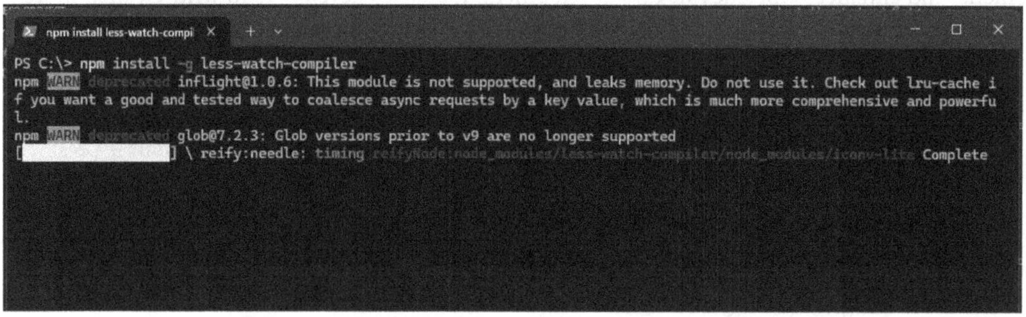

Figure 11.6 – Terminal installing less-watch-compiler

2. **Open your project.**

    Once the installation is complete, navigate to your project folder, less-project, and open it with VS Code or your preferred IDE.

3. **Organize your project structure.**

To keep your project organized, we'll modify the folder structure. Create a new folder named `less` and move the `styles.less` file into this folder. Your project structure should now look like this:

Figure 11.7 – Final project structure

4. **Configure the watcher.**

To configure `less-watch-compiler`, open your Terminal in VS Code by pressing *Ctrl + J*. Alternatively, you can use an external Terminal; just make sure you navigate to the project's folder. In the Terminal, type the following command:

```
less-watch-compiler less styles styles.less
```

This command specifies the source (`less`) and destination (`styles`) folders for the Less and CSS files, respectively. It also designates `styles.less` as the file to watch. If the last parameter is omitted, `less-watch-compiler` will watch any `.less` file inside the `less` folder.

The general syntax for this command is as follows:

```
less-watch-compiler < source-folder > <output-folder> main-file-
name
```

If everything has been set up correctly, your Terminal should display the following output:

Figure 11.8 – less-watch-compiler working in a project

5. **Test the setup.**

To ensure everything is working properly, make a small change to the `styles.less` file. For example, change `@primary-color` to `#c9041e`:

```
@primary-color: #c9041e;
```

Save the file and check the `styles` folder. You should see the updated `styles.css` file reflecting your changes:

```css
body {
  font-family: Arial, sans-serif;
  font-size: 16px;
  color: #c9041e;
}
h1 {
  color: #fb3852;
}
.btn-primary {
  background-color: #c9041e;
  border: none;
  padding: 10px;
  color: white;
}
.btn-primary:hover {
  background-color: #970317;
}
```

Open your HTML file in a web browser to verify that the new styles have been applied. Yours should look as follows:

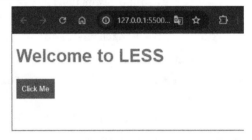

Figure 11.9 – New HTML output with automatically edited CSS

Well done! By following these steps, you just automated the process of compiling your Less files, allowing you to focus more on creating exceptional web experiences and less on manual compilation tasks.

> **Explore the less-watch-compiler documentation**
>
> For more detailed information and advanced usage, take a look at the `less-watch-compiler` documentation at `https://github.com/jonycheung/deadsimple-less-watch-compiler`. This documentation provides comprehensive guides, examples, and troubleshooting tips to help you maximize the utility of the tool and efficiently automate your Less compilation process.

Now that you've automated your Less compilation with `less-watch-compiler`, it's time to take your workflow to the next level. Introducing build tools such as Gulp and Webpack can further streamline your development process, offering powerful capabilities for task automation, module bundling, and asset optimization.

Let's explore how these tools can enhance your web development projects and improve efficiency.

Introduction to build tools – Gulp and Webpack

Build tools are software applications that automate and streamline various tasks involved in the web development process. These tools manage the compilation, optimization, and deployment of code, among other tasks.

Build tools are integral to modern web development, offering significant advantages in terms of efficiency, code optimization, and workflow enhancement. These tools manage the compilation, optimization, and deployment of code, among other tasks. They help transform source code into a production-ready application by performing tasks such as the following:

- **Compiling**: Converting code from one language into another (for example, Less/Sass into CSS or TypeScript into JavaScript)

- **Minifying**: Reducing the size of code files by removing unnecessary characters (for example, whitespace and comments) without affecting functionality

- **Bundling**: Combining multiple files into a single file to reduce the number of HTTP requests

- **Transpiling**: Converting modern JavaScript (ES6+) into a version that is compatible with older browsers

- **Linting**: Analyzing code for potential errors and enforcing coding standards

- **Live reloading**: Automatically refreshing the browser whenever changes are made to the source code

As mentioned earlier, build tools play a crucial role in modern web development. They are highly recommended for empowering web projects and benefiting developers in several ways:

- **Efficiency and productivity**: By automating repetitive tasks, build tools save developers significant time and effort. Tasks such as minifying CSS, optimizing images, or transpiling JavaScript can be automatically handled by build tools, allowing developers to focus on writing code. Automated processes ensure that every build is consistent, reducing the likelihood of human error.

- **Code optimization**: Build tools optimize code to improve performance. Minification and bundling reduce the size of files and the number of HTTP requests, which can lead to faster page load times.

  Transpiling ensures that modern JavaScript features are compatible with older browsers, broadening the reach of web applications.

- **Development workflow**: Build tools promote modular development, making it easier to manage and maintain code. Tools such as Webpack allow developers to break down code into smaller, more manageable modules. Immediate feedback through live reloading enhances the development experience, making it quicker to test and debug code changes.

- **Deployment**: Build tools prepare code for production by optimizing and packaging it, ensuring that it is efficient and ready for deployment. Many build tools can be integrated into **continuous integration/continuous deployment (CI/CD)** pipelines, automating the deployment process and ensuring that code changes are deployed quickly and reliably.

- **Scalability**: Build tools help maintain large code bases by enforcing coding standards and ensuring consistent code quality. By standardizing the build process, build tools facilitate collaboration among team members, ensuring that everyone works with the same code structure and quality standards.

Two of the most popular build tools are Gulp and Webpack. While Gulp is a task runner that automates common development tasks, Webpack is a module bundler that takes modules with dependencies and generates static assets representing those modules.

Let's explore each of these tools and learn how to enhance our development with them.

How Gulp works

Imagine that you have a web project where you need to minify your CSS, concatenate your JavaScript files, and optimize your images. Doing this manually each time you make a change would be tedious. Here's where Gulp comes in.

Gulp uses Node.js streams to process files, enabling efficient handling of large files.

With Gulp, you can set up tasks to automate these processes. For instance, you could create a `gulpfile.js` file like this:

```
const gulp = require('gulp');
const concat = require('gulp-concat');
const uglify = require('gulp-uglify');
const imagemin = require('gulp-imagemin');

// Concatenate and minify JS files
gulp.task('scripts', function () {
    return gulp.src('src/js/*.js')
        .pipe(concat('main.min.js'))
        .pipe(uglify())
        .pipe(gulp.dest('dist/js'));
});
```

```
// Optimize images
gulp.task('images', function () {
    return gulp.src('src/img/*')
        .pipe(imagemin())
        .pipe(gulp.dest('dist/img'));
});

// Watch for changes
gulp.task('watch', function () {
    gulp.watch('src/scss/*.scss', gulp.series('sass'));
    gulp.watch('src/js/*.js', gulp.series('scripts'));
    gulp.watch('src/img/*', gulp.series('images'));
});

// Default task
gulp.task(
    'default',
    gulp.series(
        'sass',
        'scripts',
        'images',
        'watch'
    )
);
```

Here's how this works:

- `gulp.task`: This defines tasks such as concatenating and minifying JavaScript (scripts) and optimizing images (images)

- `gulp.src` and `.pipe`: These methods define source files and apply transformations such as Sass compilation, JavaScript concatenation, and image optimization

- `gulp.watch`: This monitors files for changes and triggers corresponding tasks

---

Dive deeper – explore Gulp's documentation

To explore further and deepen your understanding of Gulp, we encourage you to read Gulp's documentation at `https://gulpjs.com/`. It provides comprehensive insights, examples, and best practices that will help you leverage Gulp effectively in your web development projects.

Now that we've learned how Gulp works – an automation tool that streamlines tasks such as minification, compilation, and live reloading – let's put it into practice by creating an automation task for CSS minification. Let's get coding!

Exercise 11.03 – automating the minification of CSS files with Gulp

In this exercise, you will set up a simple Gulp project to automate the minification of CSS files. Follow these steps to complete the exercise:

1. **Create a project folder.**

    Create a new folder for your project named `gulp-project`. Navigate to the folder in your Terminal or open the folder with VS Code and press *Ctrl + J* to open the local Terminal.

2. **Initialize npm.**

    Run the following command to create a `package.json` file:

    ```
    npm init -y
    ```

    You'll notice that it also creates the `node_modules` folder, which contains all libraries and plugins npm saves for running your project.

3. **Install Gulp and its plugins.**

    Now, install Gulp globally (if it's not already installed) by running the following command line in your terminal:

    ```
    npm install -g gulp-cli
    ```

    In sequence, install Gulp locally in your project with the following command:

    ```
    npm install --save-dev gulp
    ```

    Finally, install the necessary plugins for CSS minification and live reloading:

    ```
    npm install --save-dev gulp-clean-css browser-sync
    ```

4. **Create the base file structure.**

    Let's create the project structure. In the root folder (`gulp-project`), create a folder named `src`. Inside `src`, create an `index.html` file. Within `src`, create another folder named `css`, and inside `css`, create a file named `styles.css`.

    Next, create a folder named `dist` in the `gulp-project` folder. Also, create a file named `gulpfile.js` directly inside the `gulp-project` folder. Your file structure should now look like this:

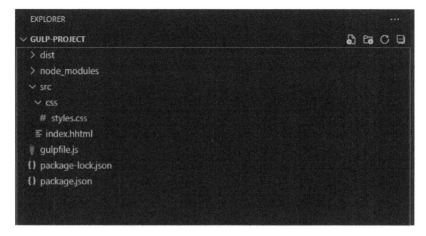

Figure 11.10 – The Gulp project's initial structure

5. **Add sample CSS and HTML.**

   Add some basic styles to `styles.css` inside the `src/css/` folder:

   ```
   body {
       font-family: Arial, sans-serif;
       background-color: #3498db;
       color: white;
       text-align: center;
       padding: 50px;
   }
   ```

   Create a simple HTML file called `index.html` in the `src/` folder:

   ```
   <!DOCTYPE html>
   <html lang="en">
   <head>
       <meta charset="UTF-8">
       <meta
           name="viewport"
           content="width=device-width,
           initial-scale=1.0">
       <title>Gulp Project</title>
       <link rel="stylesheet" href="./css/styles.css">
   </head>
   <body>
       <h1>Hello, Gulp!</h1>
   </body>
   </html>
   ```

6. **Create the Gulp configuration.**

    Open gulpfile.js and set up some tasks.

    First, import the libraries Gulp we'll be using:

    ```
    const gulp = require('gulp');
    const cleanCSS = require('gulp-clean-css');
    const browserSync = require('browser-sync').create();
    ```

    Then, create a task to minify CSS:

    ```
    gulp.task('minify-css', function() {
        return gulp.src('src/css/*.css')
            .pipe(cleanCSS())
            .pipe(gulp.dest('dist/css'))
            .pipe(browserSync.stream());
    });
    ```

    Now, create another task to copy HTML files to the dist folder:

    ```
    gulp.task('copy-html', function() {
        return gulp.src('src/*.html')
            .pipe(gulp.dest('dist'))
            .pipe(browserSync.stream());
    });
    ```

    Then, create one to initialize BrowserSync:

    ```
    gulp.task('browser-sync', function() {
        browserSync.init({
            server: {
                baseDir: './dist'
            }
        });
    });
    ```

    Create another to watch CSS and HTML files for changes:

    ```
        gulp.watch(
            'src/css/*.css',
            gulp.series('minify-css')
        );
        gulp.watch(
            'src/*.html',
            gulp.series('copy-html')
        );
    });
    ```

Finally, create a task with CSS minification and live reloading:

```
gulp.task(
    'default',
    gulp.series(
        'minify-css',
        'copy-html',
        'browser-sync'
    )
);
```

The complete `gulpfile.js` file should look like this:

```
const gulp = require('gulp');
const cleanCSS = require('gulp-clean-css');
const browserSync = require('browser-sync').create();

gulp.task('minify-css', function() {
    return gulp.src('src/css/*.css')
        .pipe(cleanCSS())
        .pipe(gulp.dest('dist/css'))
        .pipe(browserSync.stream());
});

gulp.task('copy-html', function() {
    return gulp.src('src/*.html')
        .pipe(gulp.dest('dist'))
        .pipe(browserSync.stream());
});

gulp.task('browser-sync', function() {
    browserSync.init({
        server: {
            baseDir: './dist'
        }
    });

    gulp.watch(
        'src/css/*.css',
        gulp.series('minify-css')
    );
    gulp.watch(
        'src/*.html',
        gulp.series('copy-html')
    );
```

```
    });

    gulp.task(
        'default',
        gulp.series(
            'minify-css',
            'copy-html',
            'browser-sync'
        )
    );
```

7. **Run Gulp.**

    After setting everything up, in your Terminal, run the default Gulp task to start the minification process and enable live reloading:

    ```
    Gulp
    ```

    Open your web browser and navigate to `http://localhost:3000`. You should see your `index.html` page with the applied CSS styles.

    Modify `styles.css` or `index.html` and observe the changes be automatically reflected in the browser without manual refresh.

With Gulp set up and running smoothly, it's time to explore another powerful tool: Webpack. Webpack is a module bundler that manages and optimizes your project's assets, including JavaScript, CSS, images, and more. In the next section, we'll dive into how Webpack works, configure it for your project, and leverage its capabilities to enhance your build process.

How Webpack works

Webpack is another powerful build tool that uses a modular approach to handle assets and dependencies in your web projects. Similar to Gulp, Webpack automates tasks and optimizes your development workflow through its configuration file (`webpack.config.js`). Here's a basic overview of how Webpack works:

- **Module bundling**: Webpack treats all files in your project as modules, allowing you to use import and export statements to manage dependencies. It bundles these modules into a few optimized bundles that your browser can understand, reducing the number of HTTP requests.

- **Asset handling**: In addition to JavaScript, Webpack can manage other assets, such as CSS, images, fonts, and more. It uses loaders to process different file types, transforming them as needed (for example, transpiling TypeScript to JavaScript and converting Sass into CSS).

- **Configuration**: You define how Webpack processes your files in `webpack.config.js`. This file specifies entry points, output paths, loaders, plugins, and other settings crucial for your project's build process.

Here's an example of a basic `webpack.config.js` file:

```js
const path = require('path');

module.exports = {
  entry: './src/index.js',
  output: {
    path: path.resolve(__dirname, 'dist'),
    filename: 'bundle.js',
  },
  module: {
    rules: [
      {
        test: /\.js$/,
        exclude: /node_modules/,
        use: {
          loader: 'babel-loader',
          options: {
            presets: ['@babel/preset-env'],
          },
        },
      },
      {
        test: /\.css$/,
        use: ['style-loader', 'css-loader'],
      },
      {
        test: /\.(png|svg|jpg|gif)$/,
        use: ['file-loader'],
      },
    ],
  },
};
```

In summary, Webpack enhances your web development workflow by automating tasks, optimizing performance, and improving code organization. Whether you're building a small website or a complex web application, Webpack's capabilities make it an invaluable tool for modern web development.

> **Dive deeper – explore Webpack's documentation**
>
> Visit `webpack.js.org` to uncover comprehensive guides, tutorials, and best practices that will help you harness the full potential of this powerful module bundler.

Let's get familiar with Webpack by setting up our first Webpack project.

Exercise 11.04 – setting up your first Webpack project

Let's set up a basic project that bundles JavaScript files and manages static assets so that you can familiarize yourself with Webpack:

1. **Create the project's structure.**

   Let's start by creating a new folder named `webpack-project`. Inside `webpack-project`, create two folders named `src` and `dist`, respectively. Inside the `src` folder, create another two folders: `js` and `css`. Inside the `js` folder, create a file named `index.js`.

   Inside the `css` folder, create another file named `styles.css`.

   The final structure should look like this:

   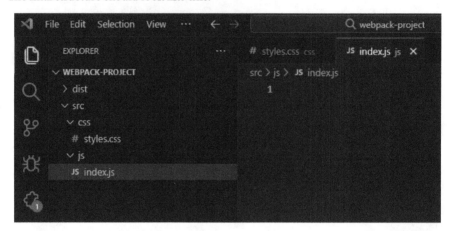

   Figure 11.11 – The webpack-project folder's structure shown in VS Code

2. **Install Webpack.**

   Open the `webpack-project` folder with your IDE and run the Command Prompt or Terminal. Make sure you're in the root folder and initialize a new npm project with the following command:

   ```
   npm init -y
   ```

   Next, install Webpack and the Webpack CLI as development dependencies by running the following command:

   ```
   npm install --save-dev webpack webpack-cli
   ```

   Also, install the plugins we're going to use in this project:

   ```
   npm install --save-dev style-loader css-loader file-loader html-
   webpack-plugin
   ```

Here, `style-loader` puts CSS right into the web page, `css-loader` lets you include CSS in your JavaScript, and `file-loader` helps bundle up images and fonts. Remember to always verify the plugin's security before using it in your project.

3. **Create base code.**

After installing the dependencies, let's create some base code to run this project. Paste the following inside the `src/js/index.js` file:

```
import '../css/styles.css';

const message = 'Hello, World!';
const h1 = document.createElement('h1');
h1.textContent = message;
document.body.appendChild(h1);
```

Then, paste the following inside the `src/css/styles.css` file:

```
body {
background-color: #F26624;
color: white;
font-family: Arial, sans-serif;
text-align: center;
margin-top: 50px;
}
```

4. **Create an HTML template.**

Create a file named `index.html` inside the `src` folder and place the following code so that it can be used as a template:

```
<!DOCTYPE html>
<html lang="en">
<head>
  <meta charset="UTF-8">
  <meta
    name="viewport"
    content="width=device-width,
    initial-scale=1.0">
  <title>Webpack Exercise</title>
</head>
<body>
</body>
</html>
```

5. **Configure Webpack.**

   In the root of your project, create a file named `webpack.config.js` and add the following configuration:

   ```js
   const path = require('path');
   const HtmlWebpackPlugin =
     require('html-webpack-plugin');

   module.exports = {
     entry: './src/js/index.js',
     output: {
       filename: 'main.js',
       path: path.resolve(__dirname, 'dist'),
     },
     module: {
       rules: [
         {
           test: /\.css$/,
           use: ['style-loader', 'css-loader'],
         },
       ],
     },
     plugins: [
       new HtmlWebpackPlugin({
         template: './src/index.html'
       }),
     ],
   };};
   ```

6. **Configure the build script.**

   We're going to run Webpack as a CLI command, so let's configure how we're going to call it using the **scripts** values inside `package.json`:

   ```json
   "scripts": {
     "build": "webpack"
   }
   ```

   Notice that you may already have a `scripts` section in your `package.json` file. If so, simply add the following content on a new line within that section:

   ```json
   "scripts": {
   [previous content],
     "build": "webpack"
   }
   ```

   Now, we can call Webpack using the `npm run build` command. This will build our project.

7. **Verify the output.**

To verify the resulting output, run the `build` command in your Terminal with `npm run build`. Webpack will create the output files inside the `dist` folder. Open the `index.html` file within the `dist` folder to see the rendered output in your browser. Additionally, examine how Webpack minifies the code and handles JavaScript by inspecting the generated files:

Figure 11.12 – Output of the project with Webpack

Well done! You've just set up your first Webpack project and gained hands-on experience with configuring Webpack, managing JavaScript and CSS assets, and automating the build process.

As an extra challenge, you can modify the `index.js` file to add an HTML element to the body using JavaScript or add additional styles to the `style.css` file and verify that the changes are reflected when you rebuild the project.

Gulp or Webpack? Which one should you choose?

Webpack and Gulp are both build tools that serve overlapping purposes in web development, but they approach tasks in different ways and are suited for different types of workflows. Here are some similarities they have:

- Both Webpack and Gulp automate tasks that would otherwise be tedious to do manually, such as minifying CSS/JS, optimizing images, and more
- Both tools use plugins (Gulp) or loaders (Webpack) to extend functionality and process different types of files, such as Sass, TypeScript, or images
- They improve the development workflow by enabling tasks such as live reloading, transpiling modern JavaScript, and optimizing assets for production

When it comes to differences, Gulp is more task-oriented and flexible in terms of the tasks you can perform and how you chain them together. It's often used for simpler tasks or when you need more control over the build process. Gulp is suitable for a wide range of tasks beyond JavaScript bundling, such as copying files, compiling Sass, running tests, or any task that can be automated using streams.

Webpack focuses heavily on module bundling and dependency management. It treats your entire project as a dependency graph and bundles assets accordingly. It's highly suited for complex applications with many dependencies. Webpack is ideal for modern JavaScript applications using frameworks that rely heavily on module-based architecture and code splitting, such as React, Vue.js, or Angular.

While both Webpack and Gulp are build tools that automate similar tasks in web development, they differ in their philosophy, approach, and typical use cases. Depending on your project's requirements and your team's preferences, you may choose one or both tools to optimize your development workflow effectively.

How to configure Gulp and Webpack for different environments (development and production)

Setting up different environments ensures that your application is tested thoroughly, optimized for performance, deployed securely, and developed efficiently from the initial stages to production release. It's a fundamental practice in modern web development that helps maintain high-quality standards and user satisfaction.

Configuring build tools to accommodate various environments such as development and production is essential for optimizing your web development workflow. Let's explore how to set up Webpack effectively to meet the demands of different environments.

Setting up different environments with Webpack

Webpack provides a simple yet effective way to specify the environment mode directly in the `webpack.config.js` file. By setting the **mode** configuration option, Webpack automatically applies various built-in optimizations tailored to the specified environment.

Here's how you can set the `mode` option:

```
module.exports = {
  // other configurations...
  mode: 'development',
};
```

Webpack accepts three values for the `mode` option:

- `mode: 'development'`: Sets `process.env.NODE_ENV` to `development` and enables useful names for modules and chunks, making debugging easier

- `mode: 'production'`: Sets `process.env.NODE_ENV` to `production`, applies several optimizations, such as deterministic mangled names for modules and chunks, and includes plugins such as `FlagDependencyUsagePlugin`, `FlagIncludedChunksPlugin`, `ModuleConcatenationPlugin`, `NoEmitOnErrorsPlugin`, and `TerserPlugin`

- `none`: Opts out of any default optimization options, providing a clean slate for custom configurations

Sometimes, you might want to adjust the Webpack configuration dynamically based on the environment. To achieve this, you can export a function instead of an object from `webpack.config.js`. Here's an example:

```
var config = {
  entry: './app.js',
  // other common configurations...
};

module.exports = (env, argv) => {
  if (argv.mode === 'development') {
    config.devtool = 'source-map';  // Enables source maps
                                    // for easier debugging
                                    // in development
  }

  if (argv.mode === 'production') {
    // Production-specific settings
    // For instance, you could add minification plugins
    // here
  }

  return config;
};
```

By leveraging the `mode` configuration option in Webpack, you can easily tailor your build process for different environments. Whether you're working in a development setting with comprehensive debugging tools or in a production environment where performance optimizations are critical, Webpack provides you with the flexibility and power to manage your configurations efficiently.

Setting up different environments with Gulp

Gulp offers a wide array of plugins that cater to different environments, each with unique features, advantages, and limitations. To find the best plugins for your specific needs, we recommend exploring the plugin section on Gulp's official website. This will help you identify the most suitable options for your projects.

As we've explored, both Gulp and Webpack are powerful tools for enhancing our applications. It's equally important to understand techniques for optimizing their use to get the most out of these tools. In the upcoming section, we'll provide a brief overview of various techniques for optimizing your build process and enhancing your application's speed and efficiency. These strategies will help you leverage Gulp and Webpack more effectively, ensuring that your projects run smoothly and perform at their best.

Techniques for improving performance with build tools

Improving website performance is the focal point of this chapter. We explored how CSS preprocessors such as Sass or Less facilitate modular and organized CSS development. Additionally, we discussed how build tools such as Webpack or Gulp automate tasks such as concatenation and minification, optimizing the size and load times of web files.

To summarize, we'll look at the primary techniques for leveraging these tools to enhance website performance.

Modular CSS development

Preprocessors such as Less and Sass allow developers to write CSS in a more modular and organized manner using features such as variables, mixins, and nested rules. This modular approach reduces redundancy and promotes code reusability, leading to more efficient CSS stylesheets.

By using variables for colors, fonts, and other style properties, preprocessors help eliminate repetitive code. This not only improves maintainability but also reduces the overall size of CSS files, contributing to faster loading times.

Minifying CSS and JavaScript

Minification significantly enhances website speed by reducing file sizes through the removal of unnecessary characters, whitespace, and comments from CSS and JavaScript files.

Both Gulp and Webpack have plugins and modules (for example, `gulp-uglify` and `gulp-clean-css` for Gulp, and `terser-webpack-plugin` and `css-minimizer-webpack-plugin` for Webpack) that automate minification and concatenation tasks during the build process.

Bundle and code splitting

Bundling groups related files into a single file (bundle) reduces the number of server requests and improves load times. Code splitting further enhances this by breaking down bundles into smaller chunks that load only when needed, reducing initial load times.

Webpack excels in bundle and code splitting configurations (`SplitChunksPlugin` and dynamic imports), allowing developers to optimize how assets are bundled and loaded based on application needs.

Optimizing images and assets

Compressing and optimizing images reduces file sizes while maintaining visual quality. Using modern image formats such as WebP can further reduce image sizes. Additionally, optimizing other assets (fonts, icons, and so on) through compression and bundling helps streamline page loading.

Gulp and Webpack provide plugins (`gulp-imagemin` and `image-webpack-loader`) that automate image and asset optimization tasks, ensuring efficient handling of resources during build processes.

By implementing these techniques with CSS preprocessors and build tools, you can streamline development workflows, optimize website performance, and deliver faster, more responsive web experiences to users.

Summary

Congratulations! By incorporating the skills you've acquired in this chapter, you've gained a comprehensive understanding of how preprocessors such as Sass and efficient tooling such as Gulp enhance web development. These tools automate tasks such as compilation, minification, and optimization, significantly improving your efficiency by saving valuable time and effort. Additionally, they optimize code output, ensuring faster loading times and better overall performance for a smoother user experience.

Beyond individual productivity gains, these tools facilitate collaboration by promoting consistent coding standards and enabling seamless teamwork across development teams. Furthermore, mastering these technologies enables you to optimize web performance, particularly in loading HTML and CSS files, contributing to enhanced project quality, efficiency, and competitiveness in the digital landscape.

In the next chapter, we'll delve deeper into enhancing code quality and mastering *Strategies for Maintaining CSS Code*. This knowledge is essential for developers aiming to create lasting web solutions that are easy to update and evolve. Get ready to elevate your coding skills!

12

Strategies for Maintaining CSS Code

In today's rapidly evolving web development landscape, maintaining a clean and efficient CSS code base has become more critical than ever. Effective CSS management is not just about writing code that works; it's also about writing code that can be easily understood, maintained, and extended by others.

This ensures that future developers can step into a project with ease, continuing the work without the need for a complete overhaul. By adopting well-supported standards and methodologies, we can create a sustainable and maintainable code base that stands the test of time.

In this chapter, we're going to cover the following main topics:

- The importance of maintainable CSS
- Understanding semantic modular CSS
- What is the **Block Element Modifier** (**BEM**) standard?
- Understanding reusable and scalable CSS

By the end of this chapter, you will be equipped with the knowledge and tools to transform your CSS into a well-organized and maintainable asset for any large-scale project.

Technical requirements

The code files for this chapter can be found at `https://packt.link/7quZK`.

The importance of maintainable CSS

As projects grow in complexity, with thousands of lines of code spanning across multiple files, the need for robust strategies to manage CSS effectively becomes paramount.

Without proper organization, CSS can quickly become unwieldy, leading to issues such as code duplication, specificity conflicts, and difficulties in maintaining and scaling a project. To prevent such chaos, it is crucial to write maintainable CSS by following best practices, such as modular CSS and the BEM approach, which we'll explore in this chapter. First, let's examine what poorly maintained CSS looks like.

What unmaintainable CSS looks like

Unmaintainable CSS isn't inherently bad coding. In smaller projects, you might not feel the need for a strong, modular approach or the use of BEM to create a sophisticated architecture for your styles. However, knowing when to apply these techniques can save time in the future and ensure long-term software quality.

Let's examine a CSS file that, while not poorly written, is unmaintainable:

```
.btn {
  padding: 10px 20px;
  background-color: blue;
  border: none;
  border-radius: 5px;
  cursor: pointer;
}

.icon {
  margin-right: 5px;
}

.primary {
  background-color: blue;
  color: white;
}

.secondary {
  background-color: gray;
  color: black;
}
```

"Well, but I write CSS this way! What's wrong with that?" you might wonder. Indeed, this code works and is perfectly fine for small projects. However, as your project grows, using global CSS such as this can lead to significant maintenance challenges. This example of CSS contains several issues that make it unmaintainable. Let's break down the problems:

- **Vague naming**: Class names such as `.btn`, `.icon`, `.primary`, or `.secondary` are too generic and do not provide enough context about their purpose or where they should be used. This lack of specificity can lead to confusion and unintended styling conflicts when similar names are used in different parts of a project.

- **Global styles**: The preceding classes are applied globally, meaning that any element in the project with these class names – ranging from menu items to gallery images – will receive styles such as `background-color: blue` or `color: black`. This can lead to unintended side effects, as different components or pages may use the same class names for different purposes.

- **Redundancy and duplication**: Styles such as `.primary` or `.secondary:` for the preceding classes are redundant. They could potentially be merged or extended from a base class to avoid duplication. This also makes future updates more cumbersome because changes need to be applied in multiple places.

In summary, unmaintainable CSS often suffers from issues such as overly specific selectors and global styles that lead to naming conflicts and reduced reusability. By adopting best practices, we can create a more maintainable, scalable, and readable code base. Let's rewrite the same example to create maintainable CSS.

What maintainable CSS looks like

The preceding unmaintainable example suffers from vague naming, global styles, lack of modularity, and redundancy. By adopting a systematic approach such as BEM, we can improve maintainability through descriptive naming, scoped styles, reusability, and reduced redundancy. This makes the CSS easier to understand, manage, and scale as a project grows.

> **What is BEM?**
> **BEM** is a methodology for writing maintainable and scalable CSS, by organizing code into reusable components with clear and descriptive class names. We'll delve deeper into BEM and its benefits in the upcoming sections.

Let's look at the result:

```
/* Improved CSS with BEM */

/* Button Component */
.button {
```

```
    padding: 10px 20px;
    border: none;
    border-radius: 5px;
    cursor: pointer;
}

.button--primary {
    background-color: blue;
    color: white;
}

.button--secondary {
    background-color: gray;
    color: black;
}

/* Icon Component */
.icon {
    margin-right: 5px;
}
```

In the preceding refactor, we have incorporated the following improvements:

- **Descriptive naming**: The new class names such as `.button`, `.button--primary`, and `.button--secondary` follow the BEM convention, making it clear that the styles are related to button components. *This descriptive naming helps other developers understand the purpose and context of each style, reducing the risk of conflicts.*

- **Scoped styles**: The styles are scoped to specific components (`.button` and `.icon`), reducing the risk of global style conflicts. *This makes the CSS more modular and easier to manage.*

- **Reusability**: By defining a base class for buttons (`.button`) and extending it with modifier classes (`.button--primary` and `.button--secondary`), the styles are more reusable and maintainable. If you need to change the padding or border radius for all buttons, you only need to update the `.button` class.

- **Reduced redundancy**: The improved version eliminates redundancy by using a base class for shared styles and modifier classes for variations. *This makes code more efficient and easier to maintain.*

In summary, maintainable CSS is structured, clear, and scalable, making it easier to work with over time. Conversely, unmaintainable CSS is often disorganized, inconsistent, and difficult to manage, leading to challenges in maintaining and extending the code base.

Why you should write maintainable CSS

Writing maintainable CSS is crucial for several reasons, especially when working on large-scale projects or in a collaborative environment. Here are some compelling reasons why you should strive to write maintainable CSS:

- **Ease of collaboration**:

  - *Consistency*: Maintainable CSS follows consistent naming conventions and patterns, making it easier for multiple developers to understand and work with code

  - *Readability*: Well-structured and documented CSS helps team members quickly comprehend styles and their purpose, reducing onboarding time for new developers

- **Scalability**:

  - *Growth*: As projects grow, maintainable CSS ensures that adding new features or components doesn't result in a tangled mess of styles

  - *Modularity*: By using methodologies such as BEM, styles can become modular and reusable, making it easy to scale the code base

- **Debugging and maintenance**:

  - *Efficiency*: Well-organized CSS reduces the time needed to find and fix bugs. Clear structure and comments help in quickly identifying and resolving issues.

  - *Future-proofing*: Maintainable CSS is easier to update and adapt to new requirements or design changes, ensuring the longevity of a project.

- **Performance**:

  - *Optimization*: Maintainable CSS often leads to cleaner and more efficient code, which can improve page load times and overall performance

  - *Minimization of redundancy*: By avoiding duplicate styles and using variables, mixins, and functions, maintainable CSS keeps file sizes smaller and more performant

- **Consistency across projects**:

  - *Uniformity*: Adhering to a style guide or coding standards ensures uniformity across different projects, making it easier for developers to switch between projects without a steep learning curve

- **Code reusability**:

  - *Efficiency*: Reusable components save time and effort, allowing developers to leverage existing styles instead of writing new ones from scratch.

- *DRY principle*: Maintainable CSS adheres to the **don't repeat yourself** principle, reducing duplication and making the code base more efficient.

- **Improved workflow**:

  - *Development speed*: A well-maintained code base speeds up development by reducing the need for rewriting or cleaning up messy CSS

  - *Tool integration*: Maintainable CSS is more easily integrated with modern development tools and workflows, such as version control systems, build tools, and preprocessors like *Sass* and *Less*. These CSS preprocessors, which we explored in Chapter 11, enhance the flexibility and scalability of CSS by introducing features such as variables, nesting, and mixins.

- **Accessibility and search engine optimization**:

  - *Standards compliance*: Writing maintainable CSS often involves adhering to web standards and best practices, which can enhance accessibility and **SEO (Search Engine Optimization)**

  - *Semantic markup*: Clear and meaningful class names contribute to semantic HTML, improving the accessibility and SEO of a website.

- **Professionalism**

  - *Quality assurance*: Writing maintainable CSS reflects a professional approach to development, showcasing attention to detail and a commitment to quality

  - *Reputation*: Consistently delivering maintainable code enhances a developer's reputation and can lead to better career opportunities and client satisfaction

Investing time and effort in writing maintainable CSS pays off in the long run. It leads to more efficient workflows, easier collaboration, better performance, and a code base that can adapt to changing requirements with minimal friction. In essence, maintainable CSS not only makes a present project manageable but also ensures that future developers will thank you for your foresight and diligence. With maintainable CSS in mind, the next step is to explore how understanding semantic modular CSS can further enhance your code's structure and readability.

Understanding semantic modular CSS

Semantic modular CSS is about writing CSS in a way that is both **meaningful** and **modular**. This means creating styles that not only describe what elements are but also how they function within a component-based structure. *The goal is to improve the maintainability, readability, and scalability of your CSS*:

- **Semantic naming** involves using class names that clearly describe the purpose and function of an element. For example, instead of using `.blue-button`, use `.btn-primary` to describe a primary button. This makes your CSS more readable and easier to understand.

- **Modular CSS** involves breaking down styles into reusable, independent modules or components. For example, instead of styling buttons differently in various parts of your style sheet, create a single `.btn` class that can be extended with modifier classes such as `.btn-primary` or `.btn-secondary`. This approach reduces duplication and makes your styles easier to maintain.

Semantic modular CSS is a methodology that focuses on writing clean, meaningful, and reusable CSS. By dividing styles into distinct modules with clear, descriptive class names, developers can create a more organized and maintainable code base. This approach enhances readability, simplifies collaboration, and makes it easier to manage and scale styles as projects grow. Let's look at how we can develop modular CSS.

How to develop modular CSS

Modular CSS is the practice of breaking down styles into reusable components, allowing developers to avoid writing styles that are used only in a single location. Instead of creating specific styles for individual elements, modular CSS encourages the use of smaller, independent style modules that can be reused in different contexts.

To implement modular CSS, you can adopt methodologies such as **SMACSS (Scalable and Modular Architecture for CSS)**. SMACSS provides a structured approach to organizing CSS, making it easier to apply modular principles effectively. It is a methodology for organizing CSS code based on its functionality. With SMACSS, you divide your CSS into five categories – base, layout, module, state, and theme. Each category contains styles specific to its purpose. Let's see how this works in practice:

- **Base styles**: These are the default styles applied to HTML elements (e.g., `h1`, `p`, and `button`). *Base styles should be minimal and reusable across an application*:

```
/* Base styles */
body {
  font-family: Arial, sans-serif;
  line-height: 1.5;
  margin: 0;
  padding: 0;
}

h1 {
  font-size: 2em;
  margin: 0.5em 0;
}
```

- **Layout styles**: Layout styles define the structure of your pages, organizing content into sections, grids, or columns:

```
/* Layout styles */
.container {
```

```
  max-width: 1200px;
  margin: 0 auto;
  padding: 0 20px;
}

.grid {
  display: flex;
  flex-wrap: wrap;
}
```

- **Module styles**: Module styles represent reusable components, such as buttons, cards, or navigation bars. Each module should have its own class, making it easy to reuse throughout your application:

```
/* Module styles */
.button {
  padding: 10px 20px;
  border: none;
  border-radius: 5px;
  cursor: pointer;
}

.button--primary {
  background-color: blue;
  color: white;
}

.button--secondary {
  background-color: gray;
  color: black;
}
```

- **State styles**: State styles manage variations in appearance based on user interaction, such as hover effects, active states, or disabled buttons:

```
/* State styles */
.button:hover {
  opacity: 0.8;
}

.button:disabled {
  background-color: lightgray;
  cursor: not-allowed;
}
```

- **Theme styles**: Theme styles can be used to define different visual themes or variations for your application, enabling easy customization without altering the base modules:

```
/* Theme styles */
.theme-dark .button {
  background-color: black;
  color: white;
}

.theme-light .button {
  background-color: white;
  color: black;
}
```

By organizing your CSS in this manner, you can create a more maintainable and scalable code base, where each file or directory contains styles specific to its category, making it easier to manage and update.

How to apply modular CSS

Now that we understand how to develop modular CSS, let's learn how to implement it. Here are the key steps:

1. *Start by analyzing your existing styles* and breaking them down into smaller, manageable modules. Look for common patterns or styles that can be abstracted into reusable components.

2. *Structure your CSS files logically, grouping related styles together.* Consider using a folder structure that mirrors the modular approach, creating separate files for base, layout, module, state, and theme styles.

3. If you are using a JavaScript framework (such as React, Vue, or Angular), take advantage of its component-based architecture. *Style each component independently, and use modular CSS principles to define styles within those components.*

> **The golden rule for modules**
> **Modules** should focus on a single responsibility. For instance, a button module is solely responsible for styling buttons. We can have modules dedicated to components, styles, or layouts.

Exercise 12.1 – implementing semantic, modular CSS with SMACSS

Let's create a simple web page using semantic HTML and modular CSS, organized according to the SMACSS methodology. This exercise will help you understand how to break down styles into base, layout, module, state, and theme categories:

1. Start by creating a folder named `semantic-modular-css` to hold our project files. Open this folder in VS Code or your preferred IDE.

2. Next, create a file named `index.html` to contain our HTML structure. Use semantic HTML tags to create a basic web page structure, including a header, a main content area with a profile card, and a footer:

```
<!DOCTYPE html>
<html lang="en">
<head>
  <meta charset="UTF-8">
  <meta
    name="viewport"
    content="width=device-width, initial-scale=1.0"
  >
  <title>Profile Card</title>
</head>
<body class="theme-light">
  <header>
    <h1>Welcome to My Page</h1>
  </header>
  <main>
    <div class="container">
      <div class="profile-card">
        <img
          src="profile-photo.jpg"
          alt="Profile Photo"
          class="profile-card__photo"
        >
        <h2 class="profile-card__name">John Doe</h2>
        <p class="profile-card__bio">
          Web developer with a passion for creating
          beautiful and functional websites.
        </p>
        <button class="button button--primary">
          Contact
        </button>
      </div>
    </div>
  </main>
  <footer>
    <p>&copy; 2024 My Website</p>
  </footer>
</body>
</html>
```

Next, create the styling files. According to SMACSS, it's important to organize your CSS into five categories – base, layout, module, state, and theme. To do this, create a folder named `styles` inside the project's folder. Within the styles folder, create separate CSS files for each category.

3. In the `base.css` file, define default styles for HTML elements (e.g., body, headings, and paragraphs):

```css
/* Base styles */
body {
  font-family: Arial, sans-serif;
  line-height: 1.5;
  margin: 0;
  padding: 0;
}

h1 {
  font-size: 2em;
  margin: 0.5em 0;
}

p {
  font-size: 1em;
  margin: 0.5em 0;
}
```

4. In the `layout.css` file, create a flexible grid layout for the page:

```css
/* Layout styles */
.container {
  max-width: 1200px;
  margin: 0 auto;
  padding: 20px;
}

header, footer {
  text-align: center;
  padding: 20px 0;
  background-color: #f0f0f0;
}
```

5. Style the profile card component and button in `module.css`:

```css
/* Module styles */
.profile-card {
  border: 1px solid #ccc;
  border-radius: 8px;
```

```css
  padding: 16px;
  max-width: 300px;
  margin: 20px auto;
  text-align: center;
  background-color: #fff;
}

.profile-card__photo {
  width: 100px;
  height: 100px;
  border-radius: 50%;
  object-fit: cover;
  margin-bottom: 16px;
}

.profile-card__name {
  font-size: 1.5em;
  margin: 8px 0;
}

.profile-card__bio {
  font-size: 1em;
  color: #666;
}

.button {
  padding: 10px 20px;
  border: none;
  border-radius: 5px;
  cursor: pointer;
}

.button--primary {
  background-color: blue;
  color: white;
}
```

6. Add interactive state styles to `state.css`:

```css
/* State styles */
.button:hover {
  opacity: 0.8;
}

.button:disabled {
  background-color: lightgray;
```

```
    cursor: not-allowed;
  }
```

7. Implement light and dark theme styles in `theme.css`:

```css
/* Theme styles */
.theme-light {
  background-color: #f9f9f9;
  color: #333;
}

.theme-dark {
  background-color: #333;
  color: #f9f9f9;
}

.theme-dark .profile-card {
  background-color: #444;
  border-color: #555;
}

.theme-dark .button--primary {
  background-color: darkblue;
}

.theme-dark header, .theme-dark footer {
  background-color: #222;
}
```

8. To use the styles we created in the exercise, you'll need to link the CSS files in the HTML document. Make sure you have the CSS files in the `styles` folder, and add `<link>` tags in the `<head>` section of your `index.html` file to include each of the CSS files:

```html
<html lang="en">
<head>
  <meta charset="UTF-8">
  <meta
    name="viewport"
    content="width=device-width, initial-scale=1.0"
  >
  <link rel="stylesheet" href="styles/base.css">
  <link rel="stylesheet" href="styles/layout.css">
  <link rel="stylesheet" href="styles/module.css">
  <link rel="stylesheet" href="styles/state.css">
  <link rel="stylesheet" href="styles/theme.css">
</head>
[…]
```

9. The resulting project should look like this:

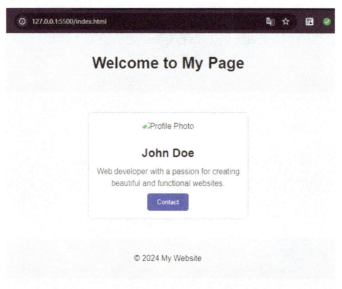

Figure 12.1 – Styled HTML with modular SMACSS CSS

10. Toggle the class on the `<body>` element between `theme-light` and `theme-dark` to switch themes, and observe the changes:

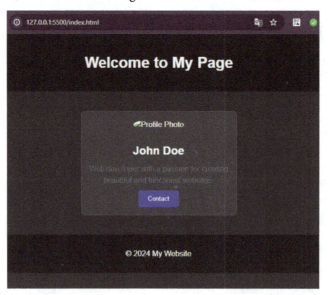

Figure 12.2 – A dark theme applied to the project

Congratulations on completing the exercise! By organizing your CSS according to the SMACSS methodology, you've made your styles more modular, maintainable, and scalable. This approach not only improves the organization of your code but also enhances its reusability across different parts of your project.

If you encounter any difficulties or want to review the exercise, you can always refer to the exercise repository to check the final result.

Modular CSS is a fundamental approach to structuring and writing CSS in a way that emphasizes reusability, clarity, and organization. As web applications grow increasingly complex, adopting a modular methodology helps developers maintain and scale their styles efficiently.

Building on the principles of modular CSS, the BEM standard offers a specific methodology to further improve reusability, clarity, and organization in your CSS code. Let's delve into BEM to understand how it can help you to structure your styles better.

What is the BEM standard?

In smaller projects, the organization of your styles isn't a major concern. You can quickly write some CSS or SASS, compile it into a single style sheet using SASS's production settings, and aggregate all the style sheets from different modules into a tidy package.

However, as projects become larger and more complex, how you organize your code is crucial for efficiency. It affects the time it takes to write code, the amount of code required, and the browser's loading time. This becomes especially important when working with teams and aiming for high performance.

One popular methodology that addresses these challenges is **BEM**. BEM is a naming convention for classes in HTML and CSS that aims to enhance code readability and maintainability by clearly defining the relationships between components. It breaks down UI components into three main parts:

- **Block**: The standalone entity that is meaningful on its own. It represents a high-level component. For example, a navigation menu (`nav`), a button (`btn`), or a form (`form`) can be a block:

  ```
  <div class="menu">...</div>
  ```

- **Element**: The part of a block that performs a specific function. It is semantically tied to its block. It is denoted with a double underscore (`__`):

  ```
  <div class="menu">
    <div class="menu__item">...</div>
  </div>
  ```

- **Modifier**: The flag on a block or element that changes its appearance or behavior. It is denoted with a double hyphen (--):

```
<div class="menu menu--large">...</div>
<div class="menu">
  <div class="menu__item menu__item--active">...</div>
</div>
```

Now that we understand what the BEM methodology is, let's apply it.

How to apply BEM

Here's a practical example that illustrates how BEM is applied:

```
<div class="card">
  <h2 class="card__title">Card Title</h2>
  <p class="card__text">
    This is some text inside the card.
  </p>
  <button class="card__button card__button--primary">
    Click Me
  </button>
</div>
```

In the style sheet, we apply the BEM strategy to clearly define blocks, elements, and modifiers:

```
.card {
  border: 1px solid #ccc;
  padding: 16px;
  border-radius: 8px;
}
.card__title {
  font-size: 1.5em;
  margin-bottom: 8px;
}
.card__text {
  font-size: 1em;
  margin-bottom: 16px;
}
.card__button {
  padding: 8px 16px;
  border: none;
  border-radius: 4px;
  cursor: pointer;
}
```

```
.card__button--primary {
  background-color: #007bff;
  color: #fff;
}
```

In the preceding example, the BEM methodology clearly organizes the CSS classes by defining the relationships between the components.

The `card` block encapsulates the entire card component, while the `card__title`, `card__text`, and `card__button` elements represent its parts.

The `card__button--primary` modifier alters the appearance of the `card__button` element, demonstrating how modifiers can be used to create variations in styling. This approach ensures that the code is easy to read, maintain, and scale. There are some additional benefits of using the BEM methodology. Let's take a look.

The benefits of BEM

BEM is particularly useful in large projects with many developers, as it provides a common language to discuss the structure and styling of components:

- **Clarity and readability**: The naming convention clearly indicates the relationship between the components, making the code easier to understand

- **Reusability**: Blocks are designed to be standalone, which encourages reusability across different parts of a project

- **Maintainability**: The modular nature of BEM helps in maintaining and updating the code base, as changes in one part of the code are less likely to affect other parts

- **Scalability**: BEM's structured approach makes it easier to scale a project, as new components can be added without disrupting the existing code base

> **Visit the documentation**
> To learn more about BEM and how to apply this methodology in your projects, check out the official documentation at `www.getbem.com`. It's a great resource to help you get started!

Exercise 12.2 – converting regular CSS into BEM

Let's convert regular CSS into BEM. This exercise will help you understand which elements should be declared as blocks, elements, or modifiers:

1. Start by creating a folder named `bem-css` to hold our project files. Open this folder in VS Code or your preferred IDE.

2. Next, create a file named `index.html` to contain our HTML structure. Use semantic HTML tags to create a basic structure and place the following code:

```html
<!DOCTYPE html>
<html lang="en">
<head>
  <meta charset="UTF-8">
  <meta
    name="viewport"
    content="width=device-width, initial-scale=1.0"
  >
<link rel="stylesheet" href="style.css">

  <title>BEM Exercise</title>
</head>
<body>
<div class="header">
  <h1 class="title">Welcome</h1>
  <nav class="nav">
    <a href="#" class="link">Home</a>
    <a href="#" class="link active">About</a>
    <a href="#" class="link">Contact</a>
  </nav>
</div>
</body>
</html>
```

3. To begin, create a `style.css` file and place the following code inside it:

```css
.header {
  background-color: #f8f9fa;
  padding: 20px;
}

.title {
  font-size: 2em;
  margin: 0;
}

.nav {
  margin-top: 10px;
}
```

```
.link {
  margin-right: 15px;
  text-decoration: none;
  color: #007bff;
}

.link.active {
  font-weight: bold;
}
```

4. This CSS will serve as the starting point for our exercise in converting to BEM methodology. When you open your HTML file, it should look like this:

Figure 12.3 – An application with regular CSS applied

5. To begin refactoring, identify the blocks, elements, and modifiers in your HTML and rename them accordingly. In this case, they should be as follows:

- **Blocks**: header and nav

- **Elements**: title (inside header) and link (inside nav)

- **Modifiers**: the active class modifies link

6. Update your HTML structure as follows:

```
<div class="header">
  <h1 class="header__title">Welcome</h1>
  <nav class="nav">
    <a href="#" class="nav__link">Home</a>
    <a href="#" class="nav__link nav__link--active">
      About
    </a>
    <a href="#" class="nav__link">Contact</a>
  </nav>
</div>
```

7. This structure reflects the application of the BEM naming convention for better organization and clarity in your CSS.

8. Let's now refactor the `style.css` file:

```css
.header {
  background-color: #f8f9fa;
  padding: 20px;
}

.header__title {
  font-size: 2em;
  margin: 0;
}

.nav {
  margin-top: 10px;
}

.nav__link {
  margin-right: 15px;
  text-decoration: none;
  color: #007bff;
}

.nav__link--active {
  font-weight: bold;
}
```

After refactoring this CSS to the BEM methodology, you'll notice that while the CSS structure improves, the user interface remains as intended.

Well done on successfully applying the BEM methodology to organize your CSS! By practicing with more complex examples, you'll further enhance your CSS development skills.

Combining strategies for optimal CSS

Combining BEM with SMACSS allows you to create modular, semantic CSS that is scalable, maintainable, and easy to understand. This approach ensures that your style sheets are structured for efficient development and collaboration.

Understanding reusable and scalable CSS

CSS has evolved significantly since its inception in the late 1990s. Originally designed to add styling to web documents, CSS has grown into a powerful language for designing complex web layouts and applications. As websites and applications became more complex, CSS had to evolve to meet new demands. Tools such as **LESS** and **Sass** extended native CSS capabilities, introducing features such as variables and `calc` functions, which greatly improved the developer experience.

Various patterns for scaling CSS emerged, aiming to balance maintenance, performance, and readability. These patterns are often referred to as **CSS architectures**, with notable examples including BEM and SMACSS.

With the rise of **SPA (Single-Page Applications)** and component-driven development, new approaches to CSS were required. Managing CSS became more challenging because components now load asynchronously, with no guarantees on source order. How do we develop reusable and scalable CSS? Let's see.

Strategies for reusable and scalable CSS

Reusable and scalable CSS refers to writing styles that can be efficiently reused across different parts of a website or application while also being adaptable to varying screen sizes and devices. Here are the key strategies to achieve this:

- **Inline styles**: In the context of components, inline styles don't face the original problem of massive duplication because they are encapsulated within the component. This encapsulation allows for the safe addition and modification of CSS in components. However, inline styles lack access to more powerful CSS features such as pseudo-selectors and media queries. Additionally, leveraging shared design tokens, caching, static analysis, and preprocessing is challenging with inline styles. Here's an example of inline styles in a JavaScript module:

```
const buttonStyle = {
  backgroundColor: 'blue',
  color: 'white',
  padding: '10px 20px',
  borderRadius: '5px'
};

const Button = () => (
  <button style={buttonStyle}>Click Me</button>
);
```

- **CSS-in-JS**: In the early days of React, Facebook introduced a JavaScript-driven approach to CSS that looked similar to inline styles but with access to style sheet power. This approach led to the proliferation of open source libraries such as **styled-components** and **Emotion**. These libraries solved many problems associated with vanilla CSS in large projects by using components, making it easy to work with dynamic values from JavaScript. However, they introduced performance issues, such as server-side rendering inefficiencies, caching problems, and client runtime costs, leading to slow app startup times.

  A more recent wave of CSS-in-JS libraries aims to combine the best of the developer experience without the runtime cost. Tools such as **Vanilla Extract**, **Linaria**, and **Compiled** extract style sheets from components during a compile step, shifting much of the runtime workload from a user's browser to compile time. Here's an example of a styled-components file:

  ```
  import styled from 'styled-components';

  const Button = styled.button`
    background-color: blue;
    color: white;
    padding: 10px 20px;
    border-radius: 5px;
  `;

  const App = () => (
    <Button>Click Me</Button>
  );
  ```

- **CSS modules**: CSS modules strike a balance between writing regular CSS (or Sass) and achieving scalability. They allow developers to use the full power of CSS without worrying about styles bleeding across components, keeping styles localized within a component directory. CSS modules do not tie CSS to a particular view library, making them a great alternative to **CSS-in-JS** libraries. Although they depend on a bundler such as **Webpack** to ensure that selectors are scoped, CSS modules are a middle ground between traditional CSS and fully component-centric approaches. They are compatible with naming conventions such as BEM. Here's an example of CSS modules applied:

  ```
  /* Button.module.css */
  .button {
    background-color: blue;
    color: white;
    padding: 10px 20px;
    border-radius: 5px;
  }
  ```

Here's the JSX file:

```
import styles from './Button.module.css';

const Button = () => (
  <button className={styles.button}>Click Me</button>
);
```

- **Utility-first CSS**: Popularized by frameworks such as **Tailwind** CSS, *utility-first CSS focuses on creating small, single-purpose utility classes that can be combined to create complex designs.* This approach promotes scalability by reducing the need to write custom CSS. According to the State of CSS, Tailwind CSS is one of the most popular implementations of this architecture today. Here's an example of Tailwind applied:

```
<button
  class="bg-blue-500 text-white py-2 px-4 rounded">
  Click Me
</button>
```

The preceding Tailwind declaration represents the following style:

```
button {
  background-color: #3b82f6;
  color: #ffffff;
  padding-top: 0.5rem;
  padding-bottom: 0.5rem;
  padding-left: 1rem;
  padding-right: 1rem;
  border-radius: 0.25rem;
}
```

Once we've learned the importance of utility-first CSS, let's dive deeper into the way it works with Tailwind next.

The implementation of utility-first CSS with Tailwind

Utility-first CSS involves creating small, reusable classes that apply a single CSS property. Instead of writing custom CSS for each component, you use these utility classes to compose your design directly in your HTML.

The benefits of utility-first CSS

Utility-first CSS offers several key benefits:

- Utility classes ensure a uniform design system across your entire application, promoting visual coherence and reducing design discrepancies

- By allowing you to apply styles directly in your HTML, this approach eliminates the need to switch contexts between HTML and CSS, streamlining the development process

- With utility-first CSS, there's a significant reduction in the need for custom CSS, leading to a cleaner, more maintainable code base that's easier to manage and update

- Additionally, utility classes can be combined in various ways to create complex designs, providing a flexible and modular approach to styling that adapts easily to different requirements and changes

Tailwind CSS is one of the most popular implementations of this methodology, offering a comprehensive set of utilities that cover a wide range of styling needs. Let's understand how it works.

> **Check out Tailwind CSS documentation for more information**
>
> For more information and to explore the full capabilities of utility-first CSS, check out `https://tailwindcss.com/docs`. It provides comprehensive guides, examples, and best practices to help you get the most out of Tailwind CSS in your projects.

How utility-first CSS works with Tailwind

To understand better how utility-first CSS works, let's see some examples of styling made with Tailwind and its regular CSS version:

Example 1 – button styling using traditional CSS

This is the version using traditional CSS:

```
<button class="custom-button">Click Me</button>

<style>
.custom-button {
  background-color: #3b82f6;
  color: #ffffff;
  padding: 0.5rem 1rem;
  border-radius: 0.25rem;
}
</style>
```

And here's its version with Tailwind CSS:

```
<button class="bg-blue-500 text-white py-2 px-4 rounded">Click Me</
button>
```

Example 2 – a card component with traditional CSS

This is the version using traditional CSS:

```
<div class="card">
  <h2 class="card-title">Card Title</h2>
  <p class="card-text">This is some text inside a card.</p>
</div>

<style>
.card {
  border: 1px solid #e5e7eb;
  padding: 1rem;
  border-radius: 0.5rem;
}

.card-title {
  font-size: 1.25rem;
  font-weight: 700;
}

.card-text {
  color: #4b5563;
}
</style>
```

And here's the same component styled with Tailwind:

```
<div class="border border-gray-300 p-4 rounded-lg">
  <h2 class="text-xl font-bold">Card Title</h2>
  <p class="text-gray-700">
    This is some text inside a card.
  </p>
</div>
```

Example 3 – responsive design with traditional CSS

This is the version using traditional CSS:

```
<div class="responsive-div">Responsive Content</div>

<style>
.responsive-div {
  padding: 1rem;
  background-color: #f3f4f6;
}

@media (min-width: 768px) {
  .responsive-div {
    padding: 2rem;
  }
}
</style>
```

And here's the same responsive design made with Tailwind:

```
<div class="p-4 bg-gray-100 md:p-8">
  Responsive Content
</div>
```

Utility-first CSS with Tailwind offers a robust, scalable, and maintainable approach to styling web applications. Writing CSS with Tailwind is significantly faster than creating traditional CSS, as it eliminates the need to manage and resolve issues related to naming and conflicting styles. In the next exercise, we'll put Tailwind into practice.

Exercise 12.3 – converting regular CSS into Tailwind

In this exercise, we will convert an HTML file styled with regular CSS into one styled with Tailwind CSS. For reference or to check the resulting files, visit the Git repository for this exercise at https://packt.link/7quZK:

1. First, create an index.html file with the initial code to use as a reference. You can find the source code at this link: https://packt.link/3s06S.

2. The output of this file is shown in *Figure 12.4*:

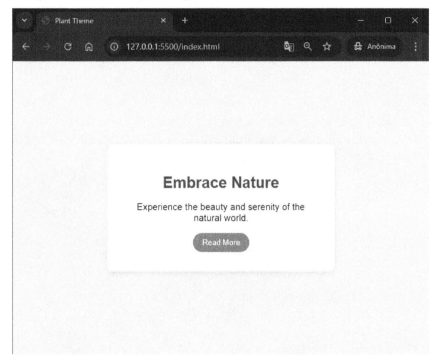

Figure 12.4 – An HTML file styled with CSS

3. Next, create a new HTML file named `index-tailwind.html` and include the bare HTML structure without any styles:

```
<!DOCTYPE html>
<html lang="en">
<head>
    <meta charset="UTF-8">
    <meta
        name="viewport"
        content="width=device-width,
                initial-scale=1.0"
    >
    <title>Plant Theme</title>
</head>
<body>
    <div class="container">
        <h1>Embrace Nature</h1>
        <p>
```

```
            Experience the beauty and serenity of the
            natural world.
        </p>
        <button>Read More</button>
    </div>
</body>
</html>
```

4. To use Tailwind directly in our HTML without installation, we'll add the **CDN** (**Content Delivery Network**) link. This method is quick and suitable for study projects. Add the CDN link just before the closing </head> tag:

```
<head>
    <meta charset="UTF-8">
    <meta
        name="viewport"
        content="width=device-width,
                initial-scale=1.0"
    >
    <title>Plant Theme</title>
    <script src="https://cdn.tailwindcss.com">
    </script>
</head>
```

5. Now, let's start styling. First, apply the background and centering styles to body:

```
<body
    class="bg-green-50
            flex items-center justify-center h-screen"
>
    ...
</body>
```

In this step, we've added a light green background color to the body. Additionally, we've centered the content both vertically and horizontally on the full screen using Tailwind's utility classes.

6. Next, style the card container itself:

```
        <div
            class="bg-white
                    text-center p-10 rounded-2xl shadow-lg
                    border border-slate-200
                    max-w-md w-full"
        >
            ...
        </div>
```

These classes style the container with a white background, rounded corners, a shadow effect, and a light green border.

7. Now, let's style the content elements (h1, p, and button):

```
<h1
    class="text-4xl font-bold text-green-800 mb-4">
    Embrace Nature
</h1>
        <p class="text-lg text-black mb-6">
            Experience the beauty and serenity of the
            natural world.
        </p>
        <button
            class="bg-green-600 text-white py-2 px-6
                rounded-full hover:bg-green-700
                transition duration-300"
        >
            Read More
        </button>
```

With these classes, we're styling the title with a bold, larger font size and a dark green color. The paragraph is styled with a medium font size and a slightly lighter green color. The button is styled with a green background, rounded edges, and a hover effect that darkens the color.

8. The resulting code should look like this:

```
<body
    class="bg-green-50 flex items-center
        justify-center h-screen"
>
    <div
        class="bg-white text-center p-10 rounded-2xl
            shadow-lg border border-slate-200
            max-w-md w-full"
    >
        <h1
            class="text-4xl font-bold text-green-800
                mb-4">
            Embrace Nature
        </h1>
        <p class="text-lg text-black mb-6">
            Experience the beauty and serenity of the
            natural world.
        </p>
        <button
```

```
                    class="bg-green-600 text-white py-2 px-6
                         rounded-full hover:bg-green-700
                         transition duration-300"
         >
              Read More
          </button>
     </div>
</body>
```

The visual output, as shown in *Figure 12.5*, is very similar to the original CSS version:

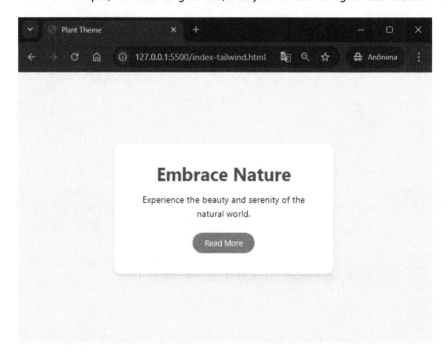

Figure 12.5 – HTML styled using Tailwind

In conclusion, converting regular CSS into Tailwind is a straightforward process that leverages utility classes to apply styles quickly and efficiently. This exercise demonstrates how Tailwind can replicate complex styles with simple, readable classes, making it a powerful tool for both rapid prototyping and production development.

Success in scaling CSS is less about rigid adherence to specific principles or best practices and more about defining needs, based on real-world constraints. It involves doing what works sustainably and performantly to get the job done.

By adopting strategies for writing reusable and scalable styles, you can create maintainable and efficient CSS code bases that contribute to a better overall user experience on the web.

Summary

In this chapter, we've explored essential strategies to maintain CSS code that are crucial for every web developer's toolkit. We began by emphasizing the importance of writing maintainable CSS, with an understanding of how thoughtful structuring and clarity in code contribute to long-term project success.

The modular CSS approach has emerged as a cornerstone, demonstrating how breaking down styles into reusable modules not only enhances code organization but also simplifies maintenance and promotes scalability. We delved into the significance of semantics, highlighting how meaningful class names and HTML structure foster code clarity and facilitate easier updates and collaboration.

The BEM methodology provides a structured framework for CSS architecture, illustrating its practical uses in managing styles across large projects. Lastly, we explored the advantages of reusable and scalable CSS, uncovering how frameworks and methodologies such as Tailwind CSS enable rapid development through utility-first approaches.

The next chapter will delve into the advancements and emerging trends in HTML and CSS. From new features and capabilities to evolving best practices, we'll explore how these technologies are shaping the future of web development. Join us as we uncover the innovations that will define the next generation of front-end design and development.

13

The Future of HTML and CSS – Advancements and Trends

As the web evolves constantly, CSS and HTML have been creating amazing new features to handle user's dynamic needs. Recently, we've seen a revolution in CSS features that empower developers to build user interfaces more efficiently and effectively. These enhancements make CSS more robust and accessible, enabling us to effortlessly create more responsive, customized, and refined websites. In this chapter, we'll cover the following main topics:

- Responsive and dynamic layouts
- Text presentation enhancements
- Color and structure customizations
- Mathematical and transformative enhancements
- Interactive and immersive elements

This chapter aims to offer an overview of these new capabilities, providing a glimpse into how they can be applied in web development projects. We'll touch on key concepts and features, giving you the foundation to experiment and explore further. For a deeper dive into these emerging tools and techniques, we encourage you to continue learning as you advance in your career.

Technical requirements

For additional coding examples and resources, check out the chapter's repository on GitHub: `https://packt.link/wpjSx`.

Responsive and dynamic layouts

In the ever-evolving landscape of web design, creating responsive and dynamic layouts has become more intuitive and powerful than ever before. The latest advancements in CSS, such as container queries, style queries, and dynamic viewport units, have revolutionized how designers and developers approach layout design. These innovations enable more granular control over elements, allowing designs that adapt not only to the viewport size but also to the context and style of their containers.

This section delves into these cutting-edge technologies, exploring how they enhance the flexibility and responsiveness of web layouts, ultimately leading to more seamless and engaging user experiences. Let's start with container queries, an innovative way to create responsive layouts.

Container queries

Container queries (`@container`) in CSS have recently become stable across all major browsers, providing a powerful new tool for responsive design. Unlike media queries, which apply styles based on the viewport size or device characteristics, container queries enable you to apply styles based on the size of an element's container.

For instance, if a container has less available space in its surrounding context, you can adjust styles accordingly—perhaps by hiding certain elements or using smaller fonts. This capability allows more granular and context-specific styling, leading to more adaptable and refined layouts. *Figure 13.1* illustrates how container queries work:

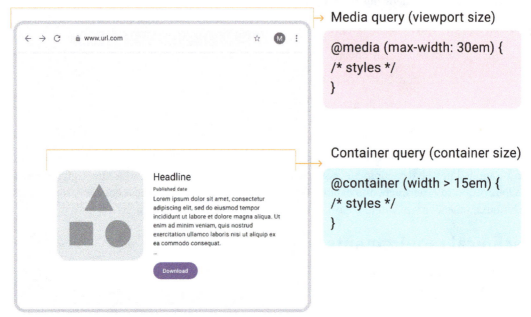

Figure 13.1 – The difference between @media queries and @container queries

To use this feature, you need to establish a containment context on an element so the browser can assess its size during rendering. This is done by setting the `container-type` property with one of the following values:

- `size`: The query will be based on both the inline and block dimensions of the container

- `inline-size`: The query will be based on the inline dimensions of the container

- `normal`: The element's styling will not be based on its size

Here's an example of a card component for a blog post that includes a title and some text:

```
<body>
    <div class="card">
        <div class="title">Blog Post Title</div>
        <div class="text">
            This is some text for the blog post.
        </div>
    </div>
</body>
```

You can establish a containment context by using the `container-type` property:

```
.card {
  border: 1px solid #ccc;
  padding: 16px;
  margin: 16px;
  container-type: inline-size;
}
```

Next, use the `@container` at-rule to define a container query. In the following example, the query will apply styles based on the size of the nearest ancestor with a containment context. Specifically, this query increases the font size of the card title if the container's width exceeds `600px`:

```
.title {
  font-size: 24px;
  margin-bottom: 8px;
}
@container (min-width: 600px) {
  .title {
  font-size: 32px;
  margin-right: 16px;
  }
}
```

You can assign a name to a containment context using the `container-name` property. You can use this name in a `@container` query to target that specific container.

The example below creates a containment context named `blogposts`:

```
.card {
  container-type: inline-size;
  container-name: blogposts;
}
```

You can then reference this containment context using the `@container` at-rule:

```
@container blogposts (min-width: 700px) {
  .title {
    font-size: 32px;
  }
}
```

Container queries represent a significant advancement in CSS, enabling us to **componentize** our layouts and move beyond relying solely on media queries for responsive design. They provide developers with the flexibility to design each component's behavior independently, which, as discussed in the previous chapter, is crucial for scaling and maintaining CSS. Additionally, the upcoming feature, **style queries**, is set to further impact layout styling significantly.

Style queries

Style queries, part of the container query specification, allow you to query the style values of a parent container based on custom properties (CSS variables). This feature provides even greater logical control over styles in CSS and enhances the separation between an application's logic and data layers from its styling. As shown in *Figure 13.2*, is possible to dynamically style blocks based on passed CSS variables:

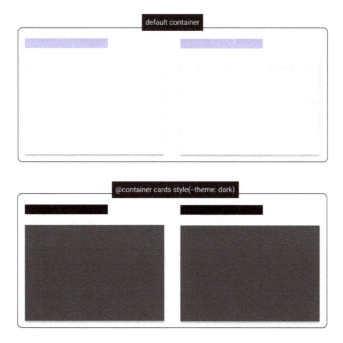

Figure 13.2 – Style queries applied to a container to alter the theme color

The `style()` notation distinguishes style queries from size queries. While support for querying regular CSS declarations such as `max-width: 100vw` is not yet available, future developments may include this capability. For now, style queries are limited to using custom properties such as the `style()` parameter, either with or without a value.

In the following example, the `.card` element's styling is based on the value of the `--season` custom property defined on the `.card-container`:

```
<ul class="card-list">
    <li class="card-container" style="--season: autumn">
        <div class="card">
        ...
        </div>
    </li>
</ul>

@container style(--season: autumn) {
    .card {
        background-color: wheat;
        border-color: brown;
    }
}
```

With style queries, you can create different styling scenarios and apply them dynamically based on user actions or data passed from the server:

```
@container style(--season: autumn) {
    .card {
        background: linear-gradient(
            -30deg,
            #ffff75,
            #fa8415
        );
    }
}


@container style(--season: summer) {
    .card {
        background: linear-gradient(
            140deg,
            #95d8ff,
            #6bffee
        );
    }
}
```

When these styles are applied to containers, you can define different styling blocks based on the variables assigned to those containers. *Figure 13.3* illustrates how these styles appear visually:

Figure 13.3 – Various styles applied through CSS variables with container queries

Style queries represent a significant shift in how we style components. They allow us to define styles based on custom properties directly within CSS and apply these styles dynamically based on data from the user or the server. Next, we'll explore dynamic viewport units, a significant addition to CSS's responsive design capabilities.

Dynamic viewport units

When using viewport units (vh for viewport height or vw for viewport width), a common issue arises on smartphones where the browser does not account for the navigation bar or on-screen buttons. As a result, elements sized at 100vh may extend beyond the visible viewport due to the discrepancy between the viewport size and the browser's UI elements. To address this, CSS has introduced several new viewport unit types to handle different viewport states:

- **Large viewport** assumes that any **user agent** (**UA**) interfaces, such as toolbars or navigation bars, are retracted. Units representing this state use the lv prefix: lvw, lvh, lvi, lvb, lvmin, and lvmax.

- **Small viewport** assumes that UA interfaces are expanded and occupy part of the screen. Units for this state use the sv prefix: svw, svh, svi, svb, svmin, and svmax.

These viewport-percentage units remain stable unless the viewport is resized. They are shown in *Figure 13.4*:

Figure 13.4 – Large and small viewport representation

Additionally, there is a dynamic viewport, which adjusts itself based on the visibility of dynamic UI elements. When dynamic toolbars are expanded, the dynamic viewport corresponds to the size of the small viewport. If they are retracted, it corresponds to the size of the large viewport, as shown in *Figure 13.5*:

Figure 13.5 – Dynamic viewport representation

Units for the dynamic viewport use the dv prefix: dvw, dvh, dvi, dvb, dvmin, and dvmax. These units are clamped between their lv and sv counterparts, providing a flexible and responsive approach to viewport sizing.

In the next section, we'll explore the latest advancements in CSS structure customizations and the innovative approaches they bring to coding with CSS.

Customizing the structure

The way we write CSS is also evolving. Recent advancements, such as cascade layers, scoped styles, nesting, and the powerful :has() selector, have significantly transformed how CSS is managed. These features empower developers to create more modular, maintainable, and intuitive styles, enabling greater control over specificity, context, and element relationships. In this section, we explore these innovative CSS capabilities, demonstrating how they streamline the styling process and open up new possibilities for crafting sophisticated and efficient web designs. Let's explore them together!

Cascade layers

CSS layers introduce a solution to the specificity problem. CSS specificity determines which styles are applied to elements by evaluating the specificity of selectors. Basic elements have lower specificity compared to classes or attributes, and IDs have the highest specificity. As you may already know, maintaining organized styles is challenging, especially with third-party code and design systems.

Using @layer, you can explicitly define the specificity of each layer, ensuring that styles do not unintentionally override one another. This approach prioritizes styles based on layer precedence rather than traditional specificity.

Consider a scenario with multiple styles for links. There are links without additional class names, one with a .nav-link class, and another with an .accent class. The CSS defines three layers: base, theme, and special:

```
@layer base {
  a {
    font-weight: bold;
    color: black; /* ignored */
  }
  .nav-link {
    color: blue; /* ignored */
  }
}

@layer theme {
  a {
    color: purple; /* styles all links */
  }
}

@layer special {
  .accent {
    color: orange;  /* styles all .accent elements */
  }
}
```

In this example, all links are either purple or orange, as *Figure 13.6* shows:

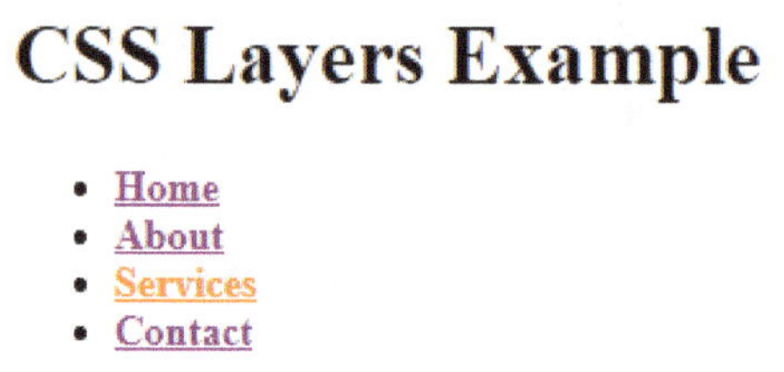

CSS Layers Example

- Home
- About
- Services
- Contact

Figure 13.6 – Styles applied using the @layers order

Even though `.nav-link` has higher selector-level specificity than a, the a `{ color: purple }` rule overrides `.nav-link { color: blue }` because the purple rule is in a higher-precedence layer. **Layer precedence takes priority over element specificity.**

You can organize layers directly on the page or at the top of a file. The order of layers is determined by the first occurrence of each layer name in your code. For instance, you can reverse the order like this:

```
@layer special, theme, base;
```

The links would appear black, and those with the `.nav-link` class would appear blue, as the `base` layer now takes precedence over the `theme` layer. *Figure 13.7* shows the output:

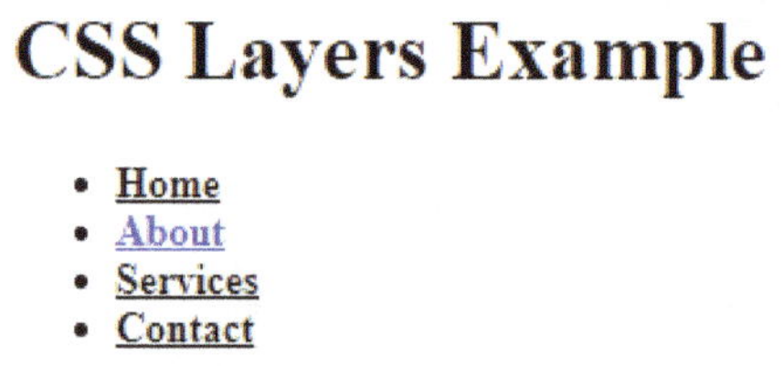

CSS Layers Example

- Home
- About
- Services
- Contact

Figure 13.7 – CSS styles applied using layers in different order

By adjusting the layer order, you can effectively manage which styles take precedence, enhancing the maintainability of your CSS.

CSS layers represent a significant advancement in CSS development, offering enhanced control over style management. To deepen your understanding and learn how to leverage CSS layers for writing more efficient and organized code, we recommend visiting the MDN web docs on CSS cascade layers: `https://developer.mozilla.org/en-US/docs/Web/CSS/@layer`. The next feature we'll see is **scoped styles**, a way to encapsulate CSS to gain more control and avoid overriding.

Scoped styles

Scoped styles allow developers to define CSS rules that apply only within a specific context or subtree. This context acts as a boundary, preventing style collisions and making it easier to manage your styles. Before scoped styles, we often relied on naming conventions or third-party libraries to achieve similar effects, but now we have a native solution. The magic behind scoped styles lies in the @scope rule. Let's break it down:

```
@scope (.card) {
  .title {
    font-weight: bold;
  }
}
```

In this example, we're scoping the .title element to live exclusively within a .card. This means that any .title inside a .card won't clash with other .title elements elsewhere on the page. The syntax of the @scope rule is as follows:

```
@scope [(<scope-start>)]? [to (<scope-end>)]? {
  /* Your stylesheet rules go here */
}
```

In this syntax, <scope-start> acts as the upper bound of the scope. In our example, .card serves as the scoping root. Everything inside it is fair game for scoped styles. <scope-ends> is optional and acts as the lower bound. If you want to narrow down the scope even further, you can specify additional elements. For instance, you might want styles to apply only to .title elements within a specific section of your .card.

Let's say we have a .media-object component with nested content, and we want to style its images and content differently:

```
@scope (.media-object) to (.content > *) {
  img {
    border-radius: 50%;
  }
  .content {
    padding: 1em;
  }
}
```

The img selector will only match image tags within a .media-object.

The .content selector will apply styles to the content within the .media-object, excluding any intervening children of the .content class.

Scoped styles provide a powerful way to encapsulate and limit the reach of your CSS rules within specific boundaries. Feel free to experiment with different scoping scenarios and create cleaner, more maintainable styles using scoped styles in your projects. Next, we'll explore CSS nesting, a highly anticipated feature that developers love and is now natively supported in CSS.

Nesting

CSS nesting, a feature beloved by developers from the days of Sass, has been one of the most requested additions to CSS for years. Now, it's finally available on the web platform, providing a more organized and succinct way to write styles. Nesting allows developers to group styles, reducing redundancy and making code easier to read and maintain.

Traditionally, you might write CSS like this:

```
.menu {}
.menu li {}
.menu li a {}
```

With nesting, you can achieve the same effect more cleanly:

```
.menu {
  li {
    a {
      /* styles for anchor elements inside list items */
    }
  }
}
```

This syntax provides a clear visual hierarchy and groups related styles together. Beyond simple structural styles, nesting also supports more advanced scenarios, such as media and container queries. Consider the following example, where a sidebar layout adapts based on its viewport's width:

```
.sidebar {
  width: 100%;
  padding: 1rem;

  @media (min-width: 600px) {
    width: 250px;
  }
}
```

In this example, when the viewport's width is 600px or more, the sidebar's width changes to 250px. The browser dynamically applies this style change when the conditions are met, making responsive design easier to manage.

Nesting extends beyond just structural elements and media queries. You can nest selectors within each other to clearly define styles for elements within a specific context:

```
.article {
  font-family: Arial, sans-serif;

  .title {
    font-size: 2rem;
    font-weight: bold;
  }
}
```

Or you can use a slightly different syntax:

```
.article {
  font-family: Arial, sans-serif;

  & .title {
    font-size: 2rem;
    font-weight: bold;
  }
}
```

In both cases, the `.title` class's styles will only apply to elements that are part of an `.article` element. This approach not only reduces the need to repeat selectors but also helps ensure that styles are more closely aligned with the HTML structure they target.

Here's another example showcasing nested styles with pseudo-classes:

```
.button {
  background-color: blue;
  color: white;

  &:hover {
    background-color: darkblue;
  }
}
```

In this case, the `:hover` pseudo-class is styled within the context of the `.button` class, changing the background color when any button using this class is hovered over.

CSS nesting streamlines the process of maintaining styles. If a component like the `.article` example is removed from the project, you can delete the entire group of styles without searching for related selectors throughout your code base.

Next, we'll learn about another groundbreaking feature that enhances CSS: the `:has()` selector.

Understanding the :has() selector

The :has() selector is one of the most powerful new features in modern CSS, offering advanced capabilities for selecting elements. This pseudo-class allows you to target a parent or a previous sibling element based on the presence of a specific reference element, using a relative selector list as its argument.

The functional :has() CSS pseudo-class selects an element if any of the provided relative selectors match at least one element within it. For instance, in the following code, :has() is used to select an <h2> element that is immediately followed by a <img> element and apply styles to the <h2> element:

```
h2:has(+ img) {
    color: red;
}
```

In the following example, you'll see how :has() can style a parent element based on the properties of its child elements:

```
.card:has(.highlight) {
    border: 2px solid #ff9800;
    background-color: #fff8e1;
}
```

The HTML structure for this example is as follows:

```
<body>
    <div class="card">
        <h2>Card Title 1</h2>
        <p>
            This is a regular paragraph without
            highlighting.
        </p>
    </div>
    <div class="card">
        <h2>Card Title 2</h2>
        <p class="highlight">
            This paragraph is highlighted with special
            styling.
        </p>
    </div>
</body>
```

Figure 13.8 illustrates the visual outcome of this styling:

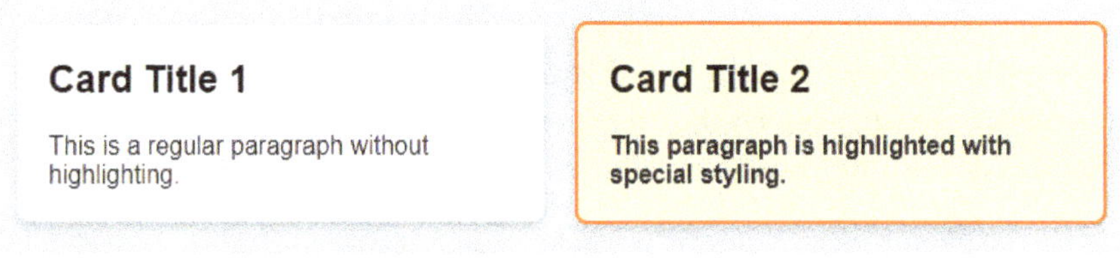

Figure 13.8 – Card styled using the :has() pseudo-class

It's important to note that the `:has()` pseudo-class cannot be nested within another `:has()`. This restriction helps prevent cyclic querying, which could arise from pseudo-elements that depend on the styles of their ancestors. Additionally, pseudo-elements cannot be used as selectors within `:has()`, nor can they serve as anchors for `:has()`.

Logical operations with :has()

The `:has()` relational selector can be utilized to check whether one or multiple conditions are met within an element's descendants. This selector accepts two conditions:

- **OR condition**: By using comma-separated values inside the `:has()` selector, you can check if any of the specified parameters exist. For instance, `x:has(a, b)` will style x if it contains either descendant a or b.

- **AND condition**: By chaining multiple `:has()` selectors together, you can check if all the specified parameters exist. For example, `x:has(a):has(b)` will style x only if it contains both descendants, a and b.

These logical operations make the `:has()` selector a versatile tool for creating more precise and dynamic CSS rules. In summary, the `:has()` selector introduces a powerful new way to conditionally style elements based on their descendants, greatly enhancing CSS's capability to create dynamic and responsive designs.

Next, we'll look at the latest advancements in CSS's mathematical and logical capabilities, including trigonometric functions and individual transform properties.

Exploring the latest CSS mathematical functions

CSS is introducing powerful mathematical and transformative capabilities that elevate web design possibilities. Among these advancements are trigonometric functions and individual transform properties, which provide developers with unprecedented control and precision in styling elements. This section delves into these cutting-edge features, exploring how they enhance the creative potential of CSS and enable more sophisticated and visually compelling designs with math and logic. Grab a calculator and let's explore!

Trigonometric functions

CSS trigonometric functions have become stable across all modern browsers, opening up exciting possibilities for creating more dynamic and organic layouts on the web. These functions allow developers to design and animate elements in ways that were previously only possible with JavaScript. The core trigonometric functions in CSS are as follows:

- `cos(angle)`: Returns the cosine of an angle, a value between -1 and 1
- `sin(angle)`: Returns the sine of an angle, a value between -1 and 1
- `tan(angle)`: Returns the tangent of an angle (ranging from negative infinity to positive infinity)

These functions can accept angles in degrees or radians, providing flexibility in how you express rotations and transformations. CSS also includes inverse trigonometric functions:

- `asin(value)`: Returns the angle corresponding to a given sine value
- `acos(value)`: Returns the angle corresponding to a given cosine value
- `atan(value)`: Returns the angle corresponding to a given tangent value
- `atan2(A, B)`: Calculates the angle between the positive x axis and the point (B, A), accepting two arguments

These functions allow more complex mathematical calculations and can help determine angles based on specific values. For example, imagine we want to draw a triangle using these functions, the CSS code will be something like this:

```css
.triangle {
--angle: 65deg; /* Angle in degrees */
--radius: 50px; /* Radius of the circle */
width: 0;
height: 0;
border-left: var(--radius) solid transparent;
border-right: var(--radius) solid transparent;
border-bottom: calc(var(--radius) * tan(var(--angle)))
    solid red;
}
```

The `sin()` and `cos()` functions are particularly useful for creating smooth animations. They oscillate between -1 and 1, which makes them perfect for cyclic movements. Trigonometric functions can add delightful motion and elegance to your designs. Whether it's a subtle animation or a complex layout, these functions are powerful tools for creative layouts.

Regarding animations, a significant upcoming change in CSS is the introduction of individual properties, which we'll explore next.

Individual transform properties

CSS now supports individual transform properties such as `translate`, `rotate`, and `scale`, significantly enhancing animation capabilities. Previously, animations were written using the combined `transform` property, like this:

```
@keyframes rotate {
0% {
  transform: rotate(0deg);
}

100% {
  transform: rotate(45deg);
}
```

With the introduction of individual properties, the same animation can now be achieved more simply:

```
@keyframes rotate {
0% {
  rotate: 0deg;
}

100% {
  rotate: 45deg;
}
```

These new properties make CSS animations easier to write and maintain, providing developers with greater control and flexibility.

As content plays a crucial role in user experience, the upcoming sections will focus on the latest features designed to enhance content handling, beginning with typography styling.

Enhancing text presentation

In the realm of web typography, recent CSS advancements have introduced exciting enhancements that improve text presentation and readability. Among these innovations are the `initial-letter` property and the `text-wrap: balance` feature, which provide designers with powerful tools to refine the visual impact and flow of textual content. This section explores these new features, demonstrating how they enhance the aesthetic and functional aspects of text on the web, enabling more engaging and polished designs. Let's start with the `initial-letter` property, which enhances control over the `:first-letter` pseudo-class.

Understanding the initial-letter property

The initial-letter CSS property is an experimental feature that allows precise styling of dropped, raised, and sunken initial letters defined by the :first-letter pseudo-class. It uses two values: the first specifies the number of lines the initial letter will span vertically, and the second represents the letter's block offset, or **sink**, determining its vertical position. Here's an example:

```
p:first-letter {
    initial-letter: 3 2;
}
```

This syntax adjusts the initial letter's size and position, as illustrated in *Figure 13.9*:

Example of initial letter styling

L orem ipsum dolor sit amet, consectetur adipiscing elit. Sed do eiusmod tempor incididunt ut labore et dolore magna aliqua. Ut enim ad minim veniam, quis nostrud exercitation ullamco laboris nisi ut aliquip ex ea commodo consequat.

Figure 13.9 – Initial letter spanning three lines in height with a two-line vertical offset

This experimental property gives developers enhanced control over the styling of the first letter in text, offering more flexibility for designing articles. Another significant advancement for frontend developers is the text-wrap: balance property, which simplifies the application of balanced text alignment.

The new CSS property text-wrap: balance

A common challenge web developers face is how to optimally break lines for titles. Designers strive for the perfect balance between lines, but achieving this across different screen sizes can be difficult. The new CSS property text-wrap: balance addresses this issue by automatically calculating and breaking title lines in an elegant and visually pleasing manner for the user. Let's understand its syntax:

```
.title-container {
    max-width: 800px;
    text-align: center;
```

```
    }
    .title {
        font-size: 1em;
    }
    .balanced-title {
        text-wrap: balance;
    }
```

As the preceding code demonstrates, simply applying the text-wrap: balance style may not yield the expected results unless the text has a defined maximum line length. This can be achieved by setting max-width on a parent container, as we've done here. The HTML structure for this example is as follows:

```
<body>
    <div class="title-container">
        <div class="title">
            <h1>
                This is a long title that needs to break
                lines without text-wrap balance
            </h1>
        </div>
        <div class="title balanced-title">
            <h1>
                This is a long title that needs to break
                lines with text-wrap balance
            </h1>
        </div>
    </div>
</body>
```

Figure 13.10 demonstrates the result: two titles, one styled without text-wrap: balance and the other with this property applied:

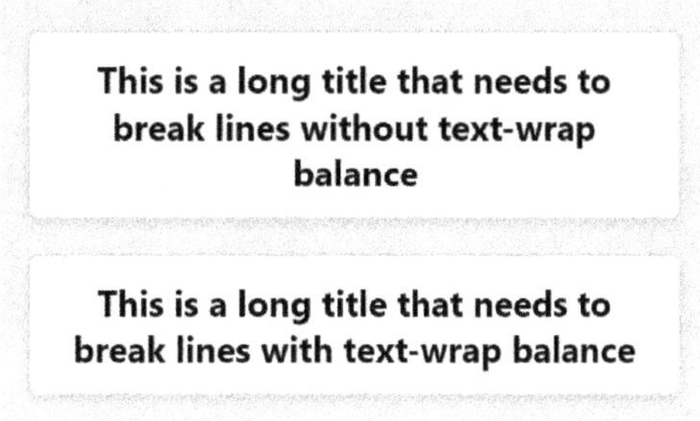

This is a long title that needs to break lines without text-wrap balance

This is a long title that needs to break lines with text-wrap balance

Figure 13.10 – Balanced title using text-wrap: balance

While this property is effective for title elements, it is not ideal for paragraphs or large blocks of text, as it may lead to performance issues. In fact, the `text-wrap: balance` property is designed to work best with text that wraps to six lines or fewer. Despite its appealing features, applying this property to all text elements is not a good idea, due to potential performance loss.

Next, we'll cover interactive and immersive elements. Let the fun begin!

Another area that has seen significant updates is user interaction. New HTML elements and CSS properties have been introduced to enhance user experience, usability, and accessibility, while also improving performance with additional native features. In the next section, we'll explore these advancements in detail.

Interactive and immersive elements

The latest advancements in CSS have introduced a suite of interactive and immersive elements that enhance user engagement and elevate web experiences. Features such as popovers, anchor positioning, select menus, discrete property transitions, and scroll-driven animations provide designers with powerful tools to create intuitive and captivating interfaces. This section explores these innovative features, demonstrating how they transform web design by enabling more responsive, engaging, and visually rich interactions. Let's start with the popover attribute.

Popover

Since April 2024, modern devices and updated browser versions have embraced the popover global attribute, revolutionizing the way we handle popovers in HTML. This attribute brings convenience, performance gains, and cleaner code—all with minimal reliance on JavaScript.

The popover global attribute is a game-changer for web developers. It automates popover behavior, reducing the need for custom JavaScript implementations. Here's how it works:

- *Automatic handling*: Popover elements remain hidden (`display: none`) until triggered by an invoking/control element. This could be a `<button>` or an `<input type="button">` with a `popovertarget` attribute. Alternatively, you can programmatically invoke a popover using `HTMLElement.showPopover()`.

- *Layer management*: When a popover opens, it appears above all other elements in the top layer. No more wrestling with z-index values! Plus, parent elements position and overflow styling won't interfere with your popover's display.

To create a popover, use the following syntax:

```
<div id="popup" popover>
  ... popover content
</div>
```

```
<button popovertarget="popup">
  Hello, I'm a popover!
</button>
```

One of the standout features of the popover global attribute is its light-dismiss behavior. Once a popover is opened, if the user clicks outside it, it will gracefully dismiss itself. No extra JavaScript is needed.

Since popovers are now natively handled, your application gains performance benefits. Next time you're building a web app, consider embracing the popover global attribute. If you want to learn more about this feature, check out the official MDN documentation at `https://developer.mozilla.org/en-US/docs/Web/HTML/Global_attributes/popover`.

The next feature is a major overhaul: the `<select>` element has been redesigned to allow greater customization. Let's dive into the details!

New select menu

The `<select>` element, as it currently exists, offers limited customization options for web developers. This often drives developers to create custom implementations, which can negatively impact performance, reliability, and accessibility compared to using native form controls.

Open UI, a community-driven initiative focused on standardizing and improving user interface components across the web, is actively working on developing a more customizable select menu. Their efforts aim to enhance the flexibility and styling capabilities of form elements while maintaining native performance and accessibility.

The future `<select>` element will maintain its current behavior unless styled with the `appearance:base-select` CSS property. In the past, suggestions for enabling additional customization included using a new tag name, adding an HTML attribute, or incorporating a child `<button>` or `<datalist>`.

Here's an example of a basic `<select>` element:

```
<select>
  <option>one</option>
  <option>two</option>
</select>
```

And here is the same `<select>` element with the new styling behavior proposed in this explainer:

```
<select style="appearance:base-select">
  <option>one</option>
  <option>two</option>
</select>
```

The new customizable select menu offers significant benefits for developers by enhancing the appearance and functionality of dropdowns. The ability to replace the default button with a custom one provides greater control over interactions and visual styling, allowing seamless integration of custom-designed buttons with the overall UI. Moreover, developers can use CSS to style the `listbox` and `button` independently, allowing for more complex UI patterns and richer user experiences.

The menu also supports the inclusion of arbitrary content within the `listbox` using `<datalist>`, facilitating intricate layouts and advanced grouping within dropdowns. Furthermore, the new select menu includes support for animations when opening and closing listboxes, providing smoother transitions and enhancing the user experience.

Overall, these enhancements empower developers to create dynamic, accessible, and visually appealing web applications using native components that align with modern design principles. To stay up to date on discussions about improving form elements, take a look at the **Open UI** explainer on the new select menu: `https://open-ui.org/components/selectlist/`.

This feature is available in Chromium-based browsers by enabling the Experimental Web Platform features in `about:flags`. Give it a try!

In the CSS realm, a new `anchor()` function has been introduced to manage element positioning. Let's explore this feature further.

Anchor positioning

The CSS Anchor Positioning CSS API allows for dynamic and flexible layout designs by tethering elements together. This feature is particularly useful in creating responsive designs where certain elements need to be aligned or positioned relative to others, regardless of changes in the viewport size or content updates.

The `anchor()` function allows precise control over the position of elements relative to their anchors. It is used within the inset property of the anchored element and helps define relationships based on edge positions. The basic syntax is as follows:

```
anchor(<anchor-element> <anchor-side>, <length-percentage>)
```

`<anchor-element>` is an optional name for the anchor element. If not provided, it defaults to the element's `position-anchor` property or its associated anchor.

`<anchor-side>` is the side of the anchor element that the positioned element aligns with. Finally, `<length-percentage>` is an optional fallback value for the function. Here's an example demonstrating how to position a tooltip anchored to a button using the `anchor()` function:

HTML

```
<body>
  <button
    id="anchorButton"
```

```
    class="anchor"
    aria-label="Button with tooltip"
  >
    Hover me!
  </button>
  <div id="tooltip" class="tooltip">
    This is a tooltip.
  </div>
</body>
```

CSS

```
.anchor {
  position: relative;
  width: 100px;
  height: 50px;
  background-color: #007bff;
  color: white;
  border: none;
  cursor: pointer;
}

.tooltip {
  position: absolute;
  inset: anchor(anchorButton end, 10px) 0 auto 0;
  background-color: #333;
  color: #fff;
  padding: 5px;
  border-radius: 3px;
  visibility: hidden;
}

.anchor:hover + .tooltip {
  visibility: visible;
}
```

In this example, button is used as the anchor, defined with an ID of anchorButton. The tooltip is positioned using the anchor() function, tethered to the end side of the anchorButton, with a 10-pixel offset. Also, the tooltip becomes visible when hovering over the button, demonstrating a basic interaction without JavaScript. *Figure 13.11* shows the visual outcome:

Figure 13.11 – Tooltip created with CSS anchor() function

This approach provides a clean and efficient way to manage element positioning in complex layouts, leveraging the power of CSS for responsive design. Next, we'll see about the new CSS discrete properties transitions.

Discrete property transitions

In the realm of web design, creating smooth transitions and animations has always been a key aspect of enhancing user experience. Traditionally, CSS animations were limited to properties that could transition smoothly over time, such as opacity or transform. However, recent advancements have introduced the ability to animate discrete CSS properties, opening new possibilities for developers.

As we discussed in *Chapter 5,* browsers are now supporting the animation of discrete properties as part of efforts to improve transitions for elements such as popovers, select menus, dialogs, and custom components.

Discrete properties are those that typically toggle between distinct values, such as moving elements to and from the top layer or transitioning display states, such as `display: none`.

Generally, these properties flip between two values at the midpoint, or 50% of the animation. This is unlike continuous properties, which transition smoothly over the duration of the animation.

To utilize these new capabilities, developers can use the `transition-behavior` property. This property allows discrete animations to be specified alongside traditional continuous transitions, providing a unified approach to animation design. Here's an example:

```
.element {
  transition: display 0.5s ease-in-out;
  transition-behavior: allow-discrete;
}
```

In this example, the `transition-behavior: allow-discrete` rule enables the discrete animation of the display property. This allows developers to incorporate complex animations seamlessly into their web projects, enhancing the interactivity and visual appeal of their interfaces.

Continuing with the theme of animations, CSS has recently introduced scroll-driven animations. Let's delve into how this feature functions.

Scroll-driven animations

Scroll-driven animations are increasingly popular in modern web design, providing engaging and interactive user experiences. These animations are directly linked to the scroll position of a container, meaning that as users scroll up or down, the animation progresses in sync with their scroll actions. Common examples include parallax effects, where background images move at different speeds than the foreground, and scroll-based reading indicators that advance as you move through content.

Traditionally, implementing scroll-driven animations involved responding to scroll events on the main thread, which often led to performance issues. This approach could cause lag and stutter, affecting the overall smoothness of the animation.

With the release of Chrome version 115, a new set of APIs—Scroll Timelines and View Timelines—has been introduced to revolutionize scroll-driven animations. These concepts integrate seamlessly with the existing **Web Animations API (WAAPI)** and CSS Animations API, enabling animations to run smoothly off the main thread. This advancement means you can achieve silky smooth, scroll-driven animations with minimal code and without compromising performance. Here's a simple example of a progress bar demonstrating how to use CSS scroll-driven animations:

```
<style>
  @keyframes grow-progress {
    from {
      transform: scaleX(0);
    }
    to {
      transform: scaleX(1);
    }
  }

  #progress {
    position: fixed;
    left: 0;
    top: 0;
    width: 100%;
    height: 1em;
    background: red;
    transform-origin: 0 50%;
    animation: grow-progress auto linear;
    animation-timeline: scroll();
  }

  .content {
    height: 200vh;
    width: 100px;
```

```
    background-color: linear-gradient(
      180deg,
      #e75656,
      #fff
    );
  }
</style>
<body>
  <div id="progress"></div>
  <div class="content"></div>
</body>
```

In this example, the animation scales an element horizontally from `scaleX(0)` (invisible) to `scaleX(1)` (fully visible) over time, using the `scroll()` timeline function, which synchronizes the animation with the scroll position of the container.

For more demos and tools to explore scroll-driven animations, visit `https://scroll-driven-animations.style/`. Scroll-driven animations are powerful tools for creating engaging and interactive web experiences. As these technologies continue to evolve, they offer a compelling way to enrich user interfaces and create memorable, immersive web experiences. Next, we'll explore the latest developments in colors and visual effects.

Exploring new colors and visual effects

As web design continues to push the boundaries of creativity, recent advancements in CSS have expanded the possibilities for colors and visual effects. Innovations such as wide-gamut color spaces, the `color-mix()` function, and view transitions empower designers to craft more vibrant, dynamic, and seamless experiences. Let's explore how these groundbreaking features revolutionize the use of color and effects in web design, paving the way for more immersive and captivating digital content.

Wide-gamut color spaces

A significant advancement in web design is the introduction of wide-gamut color spaces, which enhance how colors are represented on the web. A color gamut refers to the range of colors that a display or device can reproduce. Historically, the "millions of colors" promised by displays were constrained by the limited gamut of traditional color spaces, primarily sRGB (standard red-green-blue).

For over 25 years, sRGB has been the standard color space for CSS, defining colors through formats such as `rgb()`, `hsl()`, and `hex` codes. While sRGB has been a reliable foundation due to its widespread compatibility, it only covers about 30% of the colors perceivable by the human eye. This limitation means that even with high-definition displays, web designers couldn't fully match the vividness of colors captured in photographs or graphics.

With CSS Color Level 4, the web platform now supports a variety of new color spaces, significantly expanding the range of colors available to developers. These include REC2020, P3, XYZ, LAB, OKLAB, LCH, and OKLCH. Here are the differences between them:

- REC2020: A wide-gamut space that embraces the most vivid colors your display can handle
- P3: Popular in Apple devices, P3 extends beyond sRGB
- XYZ: A mathematically precise color space
- LAB: A perceptually uniform space that accounts for how humans perceive differences in color
- OKLAB: A perceptually uniform space optimized for digital displays
- LCH: A cylindrical representation of color (like polar coordinates)
- OKLCH: Like LCH, but tuned for digital displays

These color spaces offer a broader spectrum, enabling the use of more vivid and accurate colors that align with modern HD displays, as *Figure 13.12* shows:

Figure 13.12 – Difference between sRGB and Display P3 (source:
https://developer.apple.com/videos/play/wwdc2017/821)

In addition to expanding the color range, these new spaces provide enhanced tools for color management. For instance, while **Hue, Saturation, Lightness (HSL)** offers control over lightness, CSS now includes **Lightness, Chroma, Hue (LCH)**, which provides a more perceptual approach to lightness and color manipulation. *Figure 13.13* illustrates the broader color spectrum achievable with LCH gradients compared to HSL gradients.

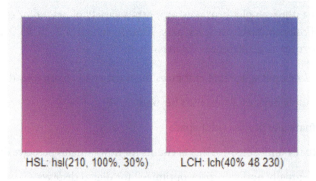

HSL: hsl(210, 100%, 30%)   LCH: lch(40% 48 230)

Figure 13.13 – The image highlights how LCH can represent a wider range of colors

As non-sRGB color spaces begin to gain traction, designers and developers will increasingly utilize these advanced tools to create richer and more vibrant color schemes.

Understanding and experimenting with these new color spaces can be a powerful addition to a designer's toolkit. Each color space has unique characteristics and use cases, allowing more precise and visually appealing designs. As these technologies evolve, embracing them will enable the creation of more immersive and dynamic web experiences. With the introduction of wide color gamuts, CSS now includes a function for more precise color mixing, which we'll examine in the following section.

The CSS color-mix() function

The color-mix() function is a powerful addition to CSS, allowing developers to blend two colors within a specified color space, providing new creative possibilities for web design. This functionality enhances color manipulation by giving precise control over how colors combine, offering a level of customization previously unattainable with traditional CSS color techniques. The color-mix() function follows a straightforward syntax:

```
color-mix(method, color1[ p1], color2[ p2])
```

Here, method is the <color-interpolation-method>, which includes the color space preceded by in, and optionally a <hue-interpolation-method>. The method defines how colors are mixed and can include various color spaces such as lch, srgb, or hsl.

The color1, color2 properties are the <color> values to be mixed. You can use color names, hex values, or any valid CSS color notation.

The properties `p1, p2` (optional) are `<percentage>` values between 0% and 100% that determine the amount of each color in the mix. The function normalizes these values to ensure they total 100%:

- If both `p1` and `p2` are omitted, the default is 50% for each color
- If `p1` is omitted, it defaults to 100% - p2
- If `p2` is omitted, it defaults to 100% - p1
- If `p1` and `p2` both equal 0%, the function is invalid
- If the sum of `p1` and `p2` does not equal 100%, the values are adjusted proportionally

Take a look at this example:

```
color-mix(in hsl longer hue, hsl(120 100% 50%) 20%, white);
```

The green hue (`hsl(120 100% 50%)`) is mixed with white, with the green contributing 20% to the final color blend. *Figure 13.14* shows a visual representation of `color-mix()` with red and blue in different color spaces:

Figure 13.14 – The color-mix() function with the same colors using different methods

The `color-mix()` function also supports custom color spaces when browsers implement the `@color-profile` feature, broadening the scope of creative design. For now, it offers extensive flexibility with existing color spaces, enabling designers to craft nuanced palettes and transitions.

By using `color-mix()`, developers can refine their color strategies, produce gradients with precise color stops, and ensure that UI components are visually cohesive across different contexts. As CSS evolves, such features continue to empower developers to push the boundaries of web aesthetics. Next, let's learn about view transitions, a new method for smoothly transitioning between views in **single-page applications (SPAs)**.

View transitions

The View Transition API empowers web developers to create seamless visual transitions between different views within a website. These transitions enhance the overall user experience, whether the site is structured as a **multi-page application (MPA)** or an SPA.

View transitions find their place in various scenarios, adding finesse to user interactions: consider a product listing page where users encounter thumbnail images. When they click on a thumbnail, it smoothly transitions into a full-size product image on the product detail page. This transition provides continuity, making the journey from overview to detail feel cohesive.

When a view transition occurs within a single document (common in SPAs), we call it a **same-document view transition**. Chrome has supported these since version 111. Here's how they work:

1. To initiate a same-document view transition, use `document.startViewTransition()`.

2. Upon calling `startViewTransition()`, the API captures the current state of the page.

3. Your callback function (provided to `startViewTransition()`) handles DOM changes.

4. After the DOM update, the API captures the new state.

Behind the scenes, the API constructs a pseudo-element tree:

```
::view-transition
└─ ::view-transition-group(root)
   └─ ::view-transition-image-pair(root)
       ├─ ::view-transition-old(root)
       └─ ::view-transition-new(root)
```

The `::view-transition` overlay sits above everything else, allowing customization with CSS. CSS animations drive the magic. Customize them using existing CSS animation properties. For instance, let's make a smooth transition with a cross-fade effect:

```css
::view-transition-old(root),
::view-transition-new(root) {
  animation-duration: 0.5s;
  animation-timing-function: ease;
}

::view-transition-old(root) {
  opacity: 1;
  animation-name: fade-out;
}

::view-transition-new(root) {
  opacity: 0;
  animation-name: fade-in;
}

@keyframes fade-out {
  from { opacity: 1; }
```

```
  to { opacity: 0; }
}

@keyframes fade-in {
  from { opacity: 0; }
  to { opacity: 1; }
}
```

The View Transitions API offers a range of features that are ready for exploration. In this section, we have begun our introduction to its capabilities, laying the groundwork for more advanced applications. As you experiment with and refine your transitions, keep in mind that each adjustment plays a crucial role in improving the overall functionality and aesthetics of your application.

Summary

It's been a long journey to get here. As we conclude this chapter on the latest advancements in HTML and CSS, it is clear that the landscape of web development is continuously evolving, bringing forth innovative tools and techniques that empower developers to create more responsive, engaging, and visually stunning web experiences.

Throughout this chapter, you've explored layout adaptability with container and style queries, enhanced efficiency using the `:has()` selector, and created layouts that fluidly respond to various conditions. You've gained skills in crafting visually balanced and dynamic text presentations, using properties such as `text-wrap: balance`, `initial-letter`, and dynamic viewport units to make your text adapt seamlessly to different environments.

You've explored the creation of interactive elements such as popovers, anchor positioning, and the new `select` element, along with designing sophisticated transitions and animations using scroll-driven animations, view transitions, and discrete property transitions.

With wide-gamut color spaces and the `color-mix()` function, you've discovered how to enrich your designs with vibrant color schemes. You've also learned to streamline your CSS using nesting, cascade layers, and scoped styles, allowing you to create more maintainable and powerful code structures. By leveraging trigonometric functions and individual transform properties, you've been equipped to design intricate animations and transformations that bring a new level of dynamism to your web projects.

Remember, this is another step in your journey of knowledge. Developers are always hungry for learning and curious about new techniques and resources that can help them create incredible things.

As you continue your journey through the web development universe, these tools will enable you to meet the growing demands of users and keep your projects at the forefront of technology.

Keep on learning, and happy coding!

Index

Symbols

A

B

www.packtpub.com

Subscribe to our online digital library for full access to over 7,000 books and videos, as well as industry leading tools to help you plan your personal development and advance your career. For more information, please visit our website.

Why subscribe?

- Spend less time learning and more time coding with practical eBooks and Videos from over 4,000 industry professionals

- Improve your learning with Skill Plans built especially for you

- Get a free eBook or video every month

- Fully searchable for easy access to vital information

- Copy and paste, print, and bookmark content

Did you know that Packt offers eBook versions of every book published, with PDF and ePub files available? You can upgrade to the eBook version at packtpub.com and as a print book customer, you are entitled to a discount on the eBook copy. Get in touch with us at customercare@packtpub.com for more details.

At www.packtpub.com, you can also read a collection of free technical articles, sign up for a range of free newsletters, and receive exclusive discounts and offers on Packt books and eBooks.

Other Books You May Enjoy

If you enjoyed this book, you may be interested in these other books by Packt:

Modern Full-Stack React Projects

Daniel Bugl

ISBN: 978-1-83763-795-9

- Implement a backend using Express and MongoDB, and unit-test it with Jest
- Deploy full-stack web apps using Docker, set up CI/CD and end-to-end tests using Playwright
- Add authentication using JSON Web Tokens (JWT)
- Create a GraphQL backend and integrate it with a frontend using Apollo Client
- Build a chat app based on event-driven architecture using Socket.IO
- Facilitate Search Engine Optimization (SEO) and implement server-side rendering
- Use Next.js, an enterprise-ready full-stack framework, with React Server Components and Server Actions

Mastering JavaScript Functional Programming - Third Edition

Federico Kereki

ISBN: 978-1-80461-013-8

- Understand when to use functional programming versus classic object-oriented programming
- Use declarative coding instead of imperative coding for clearer, more understandable code
- Know how to avoid side effects and create more reliable code with closures and immutable data
- Use recursion to help design and implement more understandable solutions to complex problems
- Define functional programing data types with or without TypeScript, add type checking, and implement immutability
- Apply advanced containers to get better structures to tackle errors and implement async programming

Packt is searching for authors like you

If you're interested in becoming an author for Packt, please visit `authors.packtpub.com` and apply today. We have worked with thousands of developers and tech professionals, just like you, to help them share their insight with the global tech community. You can make a general application, apply for a specific hot topic that we are recruiting an author for, or submit your own idea.

Share Your Thoughts

Now you've finished *Practical HTML and CSS*, we'd love to hear your thoughts! Scan the QR code below to go straight to the Amazon review page for this book and share your feedback or leave a review on the site that you purchased it from.

`https://packt.link/r/183508091X`

Your review is important to us and the tech community and will help us make sure we're delivering excellent quality content.

Download a free PDF copy of this book

Thanks for purchasing this book!

Do you like to read on the go but are unable to carry your print books everywhere?

Is your eBook purchase not compatible with the device of your choice?

Don't worry, now with every Packt book you get a DRM-free PDF version of that book at no cost.

Read anywhere, any place, on any device. Search, copy, and paste code from your favorite technical books directly into your application.

The perks don't stop there, you can get exclusive access to discounts, newsletters, and great free content in your inbox daily

Follow these simple steps to get the benefits:

1. Scan the QR code or visit the link below

https://packt.link/free-ebook/978-1-83508-091-7

2. Submit your proof of purchase
3. That's it! We'll send your free PDF and other benefits to your email directly